STREAM ECOLOGY & SELF-PURIFICATION

HOW TO ORDER THIS BOOK

BY PHONE: 800-233-9936 or 717-291-5609, 8AM–5PM Eastern Time

BY FAX: 717-295-4538

BY MAIL: Order Department
Technomic Publishing Company, Inc.
851 New Holland Avenue, Box 3535
Lancaster, PA 17604, U.S.A.

BY CREDIT CARD: American Express, VISA, MasterCard

BY WWW SITE: http://www.techpub.com

PERMISSION TO PHOTOCOPY—POLICY STATEMENT

Authorization to photocopy items for internal or personal use, or the internal or personal use of specific clients, is granted by Technomic Publishing Co., Inc. provided that the base fee of US $5.00 per copy, plus US $.25 per page is paid directly to Copyright Clearance Center, 222 Rosewood Drive, Danvers, MA 01923, USA. For those organizations that have been granted a photocopy license by CCC, a separate system of payment has been arranged. The fee code for users of the Transactional Reporting Service is 1-58716/01 $5.00 + $.25.

2ND EDITION

Stream Ecology & Self-Purification
An Introduction

FRANK R. SPELLMAN, Ph.D.
Environmental Health and Safety Manager
Hampton Roads Sanitation District
Virginia Beach, VA

JOANNE E. DRINAN
Technical Writer
Wastewater Treatment Plant Operations

Stream Ecology and Self-Purification
a**TECHNOMIC**®publication

Technomic Publishing Company, Inc.
851 New Holland Avenue, Box 3535
Lancaster, Pennsylvania 17604 U.S.A.

Copyright © 2001 by Technomic Publishing Company, Inc.
All rights reserved

No part of this publication may be reproduced, stored in a
retrieval system, or transmitted, in any form or by any means,
electronic, mechanical, photocopying, recording, or otherwise,
without the prior written permission of the publisher.

Printed in the United States of America
10 9 8 7 6 5 4 3 2 1

Main entry under title:
 Stream Ecology and Self-Purification: An Introduction, 2nd Edition

A Technomic Publishing Company book
Bibliography: p.
Includes index p. 257

Library of Congress Catalog Card No. 2001088337
ISBN No. 1-58716-086-2

For Rachel Morgan Wiedenhoeft
and
For Michael Drinan

Table of Contents

Preface   xi
Prologue   xv

1. INTRODUCTION 1

     1.1   Setting the Stage   1
     1.2   Definitions of Key Terms   3
     1.3   Levels of Organization   8
     1.4   Ecosystem   9
     1.5   Summary of Key Terms   11
     1.6   Chapter Review Questions   11

2. STREAM GENESIS AND STRUCTURE 15

     2.1   Introduction   15
     2.2   Water Flow in a Stream   16
     2.3   Stream Water Discharge   17
     2.4   Transport of Material   18
     2.5   Characteristics of Stream Channels   20
     2.6   Stream Profiles   20
     2.7   Sinuosity   20
     2.8   Bars, Riffles, and Pools   22
     2.9   The Floodplain   22
     2.10   Summary of Key Terms   23
     2.11   Chapter Review Questions   24

3. BIOGEOCHEMICAL CYCLES 25

     3.1   Nutrient Cycles   25
     3.2   Carbon Cycle   28
     3.3   Nitrogen Cycle   29
     3.4   Phosphorus Cycle   32

    3.5   Sulfur Cycle   33
    3.6   Summary of Key Terms   34
    3.7   Chapter Review Questions   35

4. ENERGY FLOW IN THE ECOSYSTEM ... 37

    4.1   Introduction   37
    4.2   Flow of Energy: The Basics   37
    4.3   Food Chain Efficiency   39
    4.4   Ecological Pyramids   41
    4.5   Productivity   42
    4.6   Productivity: The Bottom Line   44
    4.7   Summary of Key Terms   45
    4.8   Chapter Review Questions   45

5. POPULATION ECOLOGY ... 47

    5.1   The 411 on Population Ecology   47
    5.2   Population Ecology: How Is It Applied to Stream Ecology?   48
    5.3   Distribution   51
    5.4   Population Growth   52
    5.5   Population Response to Stress   55
    5.6   Population Response to Stress and Stream Ecology   61
    5.7   Summary of Key Terms   63
    5.8   Chapter Review Questions   63

6. STREAM WATER ... 65

    6.1   Where Is Earth's Water Located?   65
    6.2   Water: Earth's Blood   67
    6.3   Water, or Hydrologic, Cycle   68
    6.4   Stream Water   68
    6.5   Key Term   70
    6.6   Chapter Review Questions   70

7. FRESHWATER ECOLOGY ... 73

    7.1   Normal Stream Life   73
    7.2   Freshwater Ecology   73
    7.3   Lentic Habitat   74
    7.4   Lotic Habitat   77
    7.5   Limiting Factors   79
    7.6   Lentic Communities   86

7.7	Classification of Lakes	87
7.8	Major Differences Between Lotic and Lentic Systems	89
7.9	Summary of Key Terms	90
7.10	Chapter Review Questions	91

8. STREAM ECOLOGY 93

8.1	Introduction	93
8.2	Life Cycle of Streams	94
8.3	Unique Characteristics of Streams	94
8.4	Stream Habitats	95
8.5	Adaptations to Stream Current	96
8.6	General Stream Adaptations	98
8.7	Benthic Life: An Overview	99
8.8	Summary of Key Terms	100
8.9	Chapter Review Questions	101

9. BENTHIC MACROINVERTEBRATES 103

9.1	Benthic Macroinvertebrates	103
9.2	Benthic Macroinvertebrates: Indicator Organisms	104
9.3	Identification of Benthic Macroinvertebrates	106
9.4	Macroinvertebrates: Units of Organization	109
9.5	Variation in Diversity of Benthic Macroinvertebrate Species	110
9.6	Typical Stream Benthic Macroinvertebrates	110
9.7	Summary of Key Terms	129
9.8	Chapter Review Questions	131

10. STREAM POLLUTION 133

10.1	What Is Stream Pollution?	133
10.2	Stream Pollution Laws	134
10.3	Stream Pollutants	136
10.4	Selected Indicators of Stream Water Quality	142
10.5	Summary of Key Terms	147
10.6	Chapter Review Questions	148

11. BIOMONITORING 149

11.1	What Is Biomonitoring?	149
11.2	Biotic Index	151
11.3	Benthic Macroinvertebrate Biotic Index	153
11.4	Summary of Key Terms	154
11.5	Chapter Review Questions	154

12. SELF-PURIFICATION OF STREAMS 157

- 12.1 Balancing the "Aquarium"   157
- 12.2 Sources of Stream Pollution   159
- 12.3 Saprobity of a Stream   161
- 12.4 Organisms and Their Role in Self-Purification   168
- 12.5 Oxygen Sag (Deoxygenation)   169
- 12.6 Other Factors Affecting DO Levels in Streams   170
- 12.7 Impact of Wastewater Treatment on DO Levels in the Stream   171
- 12.8 Variables That Improve and Degrade Stream Quality   172
- 12.9 Measuring Biochemical Oxygen Demand (BOD)   180
- 12.10 Measuring Dissolved Oxygen in a Stream   180
- 12.11 Stream Purification: A Quantitative Analysis   181
- 12.12 Summary of Key Terms   186
- 12.13 Chapter Review Questions   187

13. BIOLOGICAL SAMPLING 189

- 13.1 Biological Sampling: The Nuts and Bolts of Stream Ecology   189
- 13.2 Biological Sampling: Planning   190
- 13.3 Sampling Locations (Stations)   192
- 13.4 Statistical Concepts   195
- 13.5 Sample Collection   198
- 13.6 Sampling Devices   213
- 13.7 The Bottom Line: Can We Afford Healthy Streams?   219
- 13.8 Summary of Key Terms   221
- 13.9 Chapter Review Questions   221
- 13.10 Suggested Reading   222

14. FINAL COMPREHENSIVE EXAMINATION 225

Appendix A—Answers to Chapter Review Questions   237
Appendix B—Answers to Final Comprehensive Examination   245
Related Reading   249
Index   257

Preface

THIS book deals primarily with the interrelationship (the ecology) of biota (life-forms) in a running water (stream) environment. Secondarily, this book deals with the stream's ability to self-attenuate pollution (i.e., to self-purify or cleanse itself). The bias of the book is dictated by our experience and interest and, also, by our belief that there is a great need at the present time for a basic review of ecological processes related to running waters.

The environmental challenges we face today include all of the same ones that we faced more than 30 years ago at the first Earth Day celebration in 1970. In spite of the unflagging efforts of environmental professionals (and others), environmental problems remain. Many large metropolitan areas continue to be plagued by smog, beaches are periodically polluted by oil spills, and many running waters (rivers and streams) still suffer the effects of poorly treated sewage and industrial discharges. But, considerable progress has been made. For example, many of our rivers that were once unpleasant and unhealthy are now "fishable and swimmable."

The problem with making progress in one area is that new problems are often discovered that prove to be even more intractable than those already encountered. In restoring running waters to their original pristine state, this has been found to be the case.

Those impacted by the science of stream ecology (e.g., water practitioners) must understand the effects of environmental stressors, such as toxics, on the microbiological ecosystem in running waters. Moreover, changes in that ecosystem must be measured and monitored. Thus, there is a need for water/wastewater operators and other environmental professionals to have a basic understanding of stream ecology and self-purification.

As our list of environmental concerns related to running waters grows and the very nature of the problems changes, it has been challenging to find materials suitable to train water/wastewater operators as well as students in the classroom. There has never been a shortage of well-written articles for the professional or layman, and there are now many excellent textbooks that provide cursory, nontechnical, introductory information for undergraduate students. There are also numerous scientific journals and specialized environmental texts for advanced students. However, most of these technical publications presuppose a working knowledge of fundamental stream ecology principles that a beginning student probably would not have.

The purpose of this book is to fill the gap between these general introductory science texts and the more advanced environmental science books used in graduate courses by covering the basics of ecology. Moreover, the necessary fundamental science and stream ecology principles that are generally assumed as common knowledge for an advanced undergraduate, but that may be new, or may need to be reviewed, for new water/wastewater operators or for many students in a mixed class of upper and lower division students are provided. Thus, again, there is a need for water/wastewater operators and other environmental professionals to have a basic understanding of stream ecology and self-purification. The science of stream ecology is a dynamic discipline; new scientific discoveries are made daily and new regulatory requirements are almost as frequent. Today's emphasis is placed on other aspects of stream ecology [e.g., non-point source pollution and total maximum daily load (TMDL)].

The second edition expands topics discussed in the original and provides new information. For example, following the guide of Penck: "... a case can be made for the thesis that river and hillslope processes provide the central theme of geomorphology"[1] (study of landform development under processes associated with running water). Simplifying Penck's view to make it germane to our work: the formation of a stream is important to the study of stream ecology. Therefore, a brief discussion of "idealized" stream formation (genesis of a stream) is provided in the second edition.

Basic concepts have been emphasized and definitions, descriptions, and abundant illustrations have been provided. In mathematical presentations, only a few derivations have been provided. Environmental science/engineering professors will complain of insufficient rigor. The layperson and water/wastewater operator, however, will be grateful. But, as Davis and Cornwell point out, to this question there are two sides; however important it may be to maintain a uniformly high standard in pure mathematics, the environmental student and practitioner may occasionally do well to rest content with the result of the argument.[2] We hope you are ready to rest content.

While primarily designed as an information source, and presented in simple, straightforward, easy-to-understand English, *Stream Ecology and Self-Purification* provides a look at a serious discipline, one based on years of extensive research on running waters. *Stream Ecology and Self-Purification* is suitable

---

[1]Penck, W., *Morphological Analysis of Land Forms*. London: Macmillan, p. 429, 1953.
[2]Davis, M. L. and Cornwell, D. A., *Introduction to Environmental Engineering*, 2nd ed. New York: McGraw-Hill, p. xv, 1991.

for use by both the technical practitioner in the field and by students in the classroom. Here is all the information you need to make technical and personal decisions about stream ecology.

To assure correlation to modern practice and design, illustrative problems are presented in terms of commonly used ecological parameters.

Each chapter ends with a Chapter Review Test to help evaluate mastery of the concepts presented. Before going on to the next chapter, take the Review Test, compare your answers to the key provided in Appendix A, and review the pertinent information for any problems you missed. If you miss many items, review the whole chapter. Chapter 14, the final chapter, provides a comprehensive final exam; answers are provided in Appendix B.

✓ *Note:* The symbol ✓ ("check mark") displayed in various locations throughout this text indicates or emphasizes an important point or points to study carefully.

This text is accessible to those who have no experience with stream ecology. If you work through the text systematically, an understanding of and skill in stream ecology can be acquired—adding a critical component to your professional knowledge.

FRANK R. SPELLMAN
JOANNE DRINAN

Prologue

Still Waters?

CONSIDER a river pool, isolated by fluvial processes and time from the main stream flow. To the casual visitor, who sees no more than water and rocks, it appears still, providing a kind of poetic solemnity, if only at the pool's surface.

But, a river pool is more than just a surface, and the term "still" may not correctly describe it. There are many ways to characterize "still." For sound or noise, "still" can mean inaudible, noiseless, quiet, or silent. With movement (or lack of movement), "still" can mean immobile, inert, motionless, or stationary.

The fundamental characterization of this particular pool's surface as being "still" is correct enough. Wedged in a lonely riparian corridor—formed by riverbank on one side and sandbar on the other—between a vigorous river system on its lower end and a glacier- and artesian-fed lake on its headwater end, almost entirely overhung by mossy old Sitka spruce, the surface of the large pool, at least at this particular location, is indeed still.

At close range, the pool's surface is clear, crystalline, unclouded, and transparent, and further back, it reflects, in mirror-image reversal, the forest at its edge, without the slightest ripple. The depths are hidden and unknown. A deep, slow-moving reach with muddy bottom typical of a river or stream pool is not seen; instead, the variegated tapestry of blues, greens, and blacks stitched together with threads of fine sand is seen carpeting the bottom, at least 12 feet below.

No sounds emanate from the pool. The motionless, silent water does not lap against its bank or bubble or gurgle over the gravel at its edge.

If a small stone is tossed into the river pool, the concentric circles ripple outward as the stone sinks to the pool bottom, following the laws of gravity, just as the river flows according to those same inexorable laws—downhill to the sea. The ripples die away, and the river water becomes as before, still. At the pool's edge, we look down through the massy depth to the pool bottom—the substrate.

The pool bottom is not flat or smooth, but instead is pitted and mounded occasionally with discontinuities. Gravel mounds alongside small corresponding indentations—small, shallow pits—make it apparent that gravel was removed from the indentations and piled into slightly higher mounds. From this viewpoint, the exact heights of the mounds and the depths of the indentations are difficult to judge because of visual distortion through several feet of water.

However, near the low gravel mounds (where female salmon bury their eggs, and where their young grow until they are old enough to fend for themselves) and actually through the gravel mounds, movement (water flow), an upwelling of groundwater, can be detected, explaining our ability to see the variegated color of pebbles. The mud and silt that would normally cover these pebbles have been washed away by the water's movement.

The slow, steady, inexorable flow of water in and out of the pool, along with the up-flowing of groundwater through the pool's substrate and through the salmon redds (nests), are only a few of the activities occurring within the pool, including the air above it, the vegetation surrounding it, and the damp bank and sandbar forming its sides.

If we could look at a cross-sectional slice of the pool, at the water column, the surface of the pool may carry those animals that can literally walk on water. The body of the pool may carry rotifers and protozoa and bacteria as well as many fish. Fish will also inhabit hidden areas beneath large rocks and ledges in order to escape predators. Further down is the pool bed, called the benthic zone, where the greatest numbers of creatures live, including larvae and nymphs of all sorts, worms, leeches, flatworms, clams, crayfish, dace, brook lampreys, sculpins, suckers, and water mites.

What lives in the pool bed, beneath the water, depends on whether the bed is gravelly or silty or muddy. Gravel will allow water, with its oxygen and food, to reach organisms that live underneath the pool. Many of the organisms that are found in the benthic zone may also be found underneath, in the hyporheal zone.

At the pool's outlet, and where its flow enters the main river, there are shallow places where water runs fast and is disturbed by rocks—the riffles. Only organisms that cling very well, such as net-winged midges, caddisflies, stoneflies, some mayflies, dace, and sculpins can spend much time here, and the plant life is restricted to diatoms and small algae. Riffles are a good place for mayflies, stoneflies, and caddisflies to live because they offer plenty of gravel in which to hide.

At first, we struggled to find the "proper" words to describe the river pool. Eventually, we settled on "Still Waters" because of our initial impression and because of our lack of understanding and knowledge. In reality, the pool is a dynamic habitat. Each river pool has its own biological community in which all members are interwoven in complex fashion and dependent on each other. Thus, the river pool habitat is part of a complex, dynamic ecosystem. On reflection, we realize, moreover, that anything dynamic certainly cannot be accurately characterized as "still"—including a river pool.

CHAPTER 1

Introduction

"We poison the caddis flies in a stream and the salmon runs dwindle and die. We poison the gnats in a lake and the poison travels from link to link of the food chain and soon the birds of the lake margins become victims. We spray our elms and the following springs are silent of robin song, not because we sprayed the robins directly but because the poison traveled, step by step, through the now familiar elm leaf-earthworm-robin cycle. These are matters of record, observable, part of the visible world around us. They reflect the web of life—or death—that scientists know as ecology." (Rachel Carson[3])

1.1 SETTING THE STAGE

As Rachel Carson points out, what we do to any part of our environment has an impact upon other parts. There is an interrelationship between the parts that make up our environment. Probably the best way to state this interrelationship is to define ecology. "Ecology is the science that deals with the specific interactions that exist between organisms and their living and nonliving environment."[4] Odum explains that the word "ecology" is derived from the Greek *oikos,* meaning home.[5] Therefore, ecology is the study of the relation of an organism or a group of organisms to their environment (their "home").

✓ *Note:* No ecosystem can be studied in isolation. If we were to describe ourselves, our histories, and what made us the way we are, we could not leave the world around us out of our description. So it is with streams: they are directly tied in with the world around them. They take their chemistry from the rocks and dirt beneath them as well as for a great distance around them.[6]

---

[3]Carson, R., *Silent Spring.* Boston: Houghton Mifflin, p. 189, 1962.
[4]Tomera, A. N., *Understanding Basic Ecological Concepts.* Portland, ME: J. Weston Walch, Publisher, p. 5, 1989.
[5]Odum, E. P., *Fundamentals of Ecology,* Philadelphia: Saunders College Publishing, p. 1, 1983.
[6]From Cave, C., *Ecology.* http://home.netcom.com/~cristi, p. 1, 1998.

Charles Darwin explained ecology in a famous passage in *The Origin of Species*—a passage that helped establish the science of ecology. A "web of complex relations" binds all living things in any region, Darwin writes. Adding or subtracting even a single species causes waves of change that race through the web "onwards in ever-increasing circles of complexity." The simple act of adding cats to an English village would reduce the number of field mice. Killing mice would benefit the bumblebees, whose nest and honeycombs the mice often devour. Increasing the number of bumblebees would benefit the heartsease and red clover, which are fertilized almost exclusively by bumblebees. So, adding cats to the village could end by adding flowers. For Darwin, the whole of the Galapagos archipelago argues this fundamental lesson. The volcanoes are much more diverse in their ecology than their biology. The contrast suggests that in the struggle for existence, species are shaped at least as much by the local flora and fauna as by the local soil and climate. "Why else would the plants and animals differ radically among islands that have the same geological nature, the same height, and climate?"[7]

The environment includes everything important to the organism in its surroundings. The organism's environment can be divided into four parts:

(1) Habitat and distribution—its place to live
(2) Other organisms—whether friendly or hostile
(3) Food
(4) Weather—light, moisture, temperature, soil, etc.

There are two major subdivisions of ecology: autecology and synecology. *Autecology* is the study of an individual organism or a species. It emphasizes life history, adaptations, and behavior. It is the study of communities, ecosystems, and biosphere. *Synecology* is the study of groups of organisms associated as a unit.[8]

An example of autecology would be when biologists spend their entire lifetimes studying the ecology of the salmon. Synecology, on the other hand, deals with the environmental problems caused by mankind. For example, the effects of discharging phosphorous-laden effluent into a stream involves several organisms. The activities of human beings have become a major component of many natural areas. As a result, it is important to realize that the study of ecology must involve people.

Each division of ecology has its own set of terms that is essential for communication between ecologists and those studying stream ecology and self-purification. Therefore, along with basic ecological terms, key terms that specifically

---

[7]Darwin, C., *The Origin of Species*, Suriano, G. (ed.). New York: Grammercy, p. 112, 1998.
[8]Odum, E. P., *Fundamentals of Ecology*. Philadelphia: Saunders College Publishing, p. 3, 1971.

pertain to this study of stream ecology and self-purification are defined and presented in alphabetical order in the following section.

1.2 DEFINITIONS OF KEY TERMS

To learn stream ecology, the language associated with the science must be mastered because each science has its own terms with its own accompanying definitions. Many of the terms used are unique. Others combine or "borrow" words from many different sciences. This is the case with the science of ecology. Its language has unique terms and also uses terms from engineering, biology, mathematics, hydrology, chemistry, physics, microbiology, hydraulics, geology, and other sciences.

In this section, many of the terms unique and/or common to stream ecology are identified and defined. Those terms not listed or defined in the following section are defined as they appear in the text.

- *Abdomen*—is the rear body section of some invertebrates.
- *Abiotic factor*—is the nonliving part of the environment composed of sunlight, soil, mineral elements, moisture, temperature, topography, minerals, humidity, tide, wave action, wind, and elevation.

  ✓ *Note:* Every community is influenced by a particular set of abiotic factors. While it is true that abiotic factors affect community members, it is also true that the living (biotic factors) may influence the abiotic factors. For example, the amount of water lost through the leaves of plants may add to the moisture content of the air. Also, the foliage of a forest reduces the amount of sunlight that penetrates the lower regions of the forest. The air temperature is, therefore, much lower than that in non-shaded areas.[9]

- *Aeration*—is a process whereby water and air or oxygen are mixed.
- *Aestivation*—refers to the ability of some fishes to burrow in the mud and wait out a dry period.
- *Antennae*—are flexible sensory appendages (occurring in pairs) on the heads of some invertebrates.
- *Appendages*—are any extensions or outgrowths from the body.
- *Aquatic*—means living or growing in water.
- *Autotroph (primary producer)*—is any green plant that fixes energy of sunlight to manufacture food from inorganic substances.
- *Bacteria*—are among the most common microorganisms in water. Bacteria are primitive, single-celled microorganisms (largely responsible for

[9]Tomera, A. N., *Understanding Basic Ecological Concepts*. Portland, ME: J. Weston Walch, Publisher, p. 41, 1989.

decay and decomposition of organic matter) with a variety of shapes and nutritional needs.
- *Benthic zone*—is the stream bed.
- *Biochemical Oxygen Demand (BOD)*—is a widely used parameter of organic pollution applied to wastewater and surface water that involves the measurement of the dissolved oxygen used by microorganisms in the biochemical oxidation of organic matter.
- *Biotic factor (community)*—is the living part of the environment composed of organisms that share the same area, are mutually sustaining, interdependent, and constantly fixing, utilizing, and dissipating energy.
- *Biotic index*—is a systematic survey of invertebrate aquatic organisms that is used to correlate with river quality. The diversity of species in an ecosystem is often a good indicator of the presence of pollution. The greater the diversity, the lower the degree of pollution.
- *Bristles*—are stiff hairs.
- *Bulbous*—means rounded or swollen shape.
- *Calcium carbonate*—is a white solid that occurs naturally as the mineral calcite and in limestone (also present in the shells and bones of some animals).
- *Carnivorous*—means meat eating.
- *Chlorophyll*—is a green pigment that allows plants to take in sunshine, water, and carbon dioxide, which they turn into sugar and oxygen.
- *Clarity*—is clearness or transparency.
- *Climax community*—is the terminal stage of ecological succession in an area.
- *Coil shaped*—is a form with spirals or rings around a center point.
- *Community*—in an ecological sense, includes all the populations occupying a given area.

  ✓ *Note:* In regards to community, Leopold points out "that land is a community is the basic concept of ecology, but that land is to be loved and respected is an extension of ethics. That land yields a cultural harvest is a fact long known, but latterly often forgotten."[10]

- *Competition*—is a critical factor for organisms in any community. Animals and plants must compete successfully in the community to stay alive.
- *Data*—are facts or pieces of information.
- *Decomposition*—is the breakdown of complex material into simpler substances by chemical or biological processes.
- *Digestive track*—consists of connected organs within the body through which food material passes while being broken down and absorbed.

---

[10]Leopold, A., *A Sand County Almanac: With Essays on Conservation from Round River.* New York: Ballantine, p. xix, 1970.

- *Dissolved oxygen (DO)*—is the amount of oxygen dissolved in a stream and is an indication of the degree of health of the stream and its ability to support a balanced aquatic ecosystem.
- *Distinct*—means clearly defined and easily recognized.
- *Dome shaped*—means that a form resembles half of a sphere.
- *Drainage basin (catchment area, stream's watershed)*—is the area that a stream drains.
- *Drift*—is the matter that is floating down the stream. Drift is important because it is the main source of food for many fish.
- *Ecosystem*—consists of the community and the nonliving environment functioning together as an ecological system.
- *Emigration*—is the departure of organisms from one place to take up residence in another area.
- *Eutrophication*—is the natural aging of a lake or land-locked body of water that results in organic material being produced in abundance due to a ready supply of nutrients accumulated over the years.
- *Filament*—is a very fine or thread-like fiber.
- *Fishkill*—takes place when less than 5 ppm of dissolved oxygen at 10°C (50°F) results in the death of large numbers of game fish, such as trout.
- *Floodplain*—is the area a stream floods.
- *Foraging*—means searching for food.
- *Fresh water*—is water that is not salty.
- *Fry*—are young fish from a hatchery.
- *Fungi*—is a group of organisms that lack chlorophyll and obtain nutrients from dead or living organic matter.
- *Gill tufts*—are fluffy clusters of gill filaments.
- *Gills*—are the breathing apparatus for aquatic organisms (may appear as filaments, tufts, or plates).
- *Habitat*—is a term used by ecologists to mean the place where an organism lives.
- *Heterotroph*—is any living organism that obtains energy by consuming organic substances produced by other organisms.
- *Homeostasis*—is a natural occurrence during which an individual population or an entire ecosystem regulates itself against negative factors and maintains an overall stable condition.
- *Hyporheal (hy-po-reel) zone*—is the area underground where the stream flows.
- *Immigration*—is the movement of organisms into a new area of residence.
- *Invertebrates*—are organisms without a backbone.
- *Larva/larvae (plural)*—is the juvenile form of many insects and other organisms that become different in form when they are adults.
- *Limiting factor*—is a necessary material that is in short supply, and because of the lack of it, an organism cannot reach its full potential.

- *Lobes*—are rounded projections.
- *Locomotion*—is movement from place to place.
- *Macroinvertebrates*—are animals that have no backbone and are visible without magnification.
- *Macrophytes*—are big plants.
- *Microhabitats*—describes very local habitats.
- *Monitoring*—is the repeated observation of condition, especially to detect and give warning of change.
- *Niche*—is the role that an organism plays in its natural ecosystem, including its activities, resource use, and interaction with other organisms.

  ✓ *Note:* Each species fills a number of different niches that together comprise its microhabitat and lifestyle. For instance, here are some of the niches an adult trout might fill in a stream: *predator* of drifting insects, minnows, eggs, juvenile fishes, and amphibians and *inhabitant* of a deep pool next to a run, with overhead cover.

  Species of both plants and animals usually change the niches they occupy as they age (ontogenic changes). For instance, the trout we examined above was at one time a small YOY (young-of-year) or Age I (one-year-old) trout that filled these niches in the same stream: *predator* of eggs and small creatures, like larvae, on the benthos (bottom of the stream) and *inhabitant* of a run of medium depth, with or without overhead cover.

  A species living next to the young trout, a dace, may nevertheless occupy its own special niches: *scraper and gatherer* of diatoms, algae, protozoa, and bacteria and *inhabitant* of a riffle.[11]

- *Nitrogen fixation*—is the ability of an organism to take nitrogen gas out of the air and transform it into biologically useful nitrogen.
- *Non-point pollution*—consists of sources of pollutants in the landscape, for example, agricultural runoff.
- *Nutrient*—is a material that serves as food or provides nourishment.
- *Oblong*—means elongated (stretched) from a square or circular shape.
- *Operculum*—is a lid or plate that covers the shell opening of some snails.
- *Organic*—means derived from living organisms.
- *Organic enrichment*—is the addition of nutrients from organic matter.
- *Organically polluted*—means that excess addition of organic matter has made matter unfit for living things.
- *Oval*—means shaped like an egg.

---

[11]From Cave, C., *Ecology.* http://home.netcom.com/~cristi, p. 8, 1998.

- *Oxygen*—is a colorless gas in the atmosphere that is essential for animal respiration.
- *Parasites*—are organisms that live on or in the body of different organisms from which they obtain nutrients.
- *Piscicides*—are chemicals that kill fish.
- *Platelike*—means resembling thin, flat sheets of uniform thickness.
- *Point source*—is a source of pollutants that involves discharge of pollutants from an identifiable point, such as a smokestack or sewage treatment plant.
- *Pollution*—is an adverse alteration to the environment by a pollutant.
- *Population*—is a group of organisms of a single species that inhabit a certain region at a particular time.
- *Predator*—is an organism that captures and feeds on other organisms.
- *Respiration*—is breathing, or the exchange of gases between the body and the environment.
- *Retractable*—means capable of being drawn or pulled back.
- *Riffle*—is a shallow area of a stream in which water flows rapidly over a rocky or gravelly stream bed.
- *Riparian corridor*—is the vegetation growing by the stream.
- *Riparian vegetation*—is vegetation around the stream.
- *Riprap*—are small boulders crated by truckloads and dumped along the sides of water channels.
- *Runoff*—may transmit organic waste that has been applied to a soil into surface waters (this waste may also be transmitted by rainfall and snowmelt).
- *Scavengers*—are animals that feed on dead or decaying organic matter.
- *Secrete*—means to generate and release a fluid or substance.
- *Segmented*—means divided into similar, repeated sections or units.
- *Sewage*—is the liquid waste from a community. Domestic sewage is from housing. Industrial sewage is normally from mixed industrial and residential.
- *Species*—is the basic category of biological classification consisting of similar organisms that are capable of mating and reproduction.
- *Spindly*—means slender and long in a way that suggests weakness.
- *Stream bed*—is the stream bottom or surface over which a stream flows.
- *Substrate*—the bottom of the stream.
- *Succession*—is a process that occurs subsequent to disturbance, and that involves the progressive replacement of biotic communities with others over time.
- *Symbiosis*—is a compatible association between dissimilar organisms to their mutual advantage.
- *Tapered*—means that a shape is gradually narrower or thinner toward one end.

- *Trophic level*—is the feeding position occupied by a given organism in a food chain, measured by the number of steps it is removed from the producers.
- *Umbo*—is the raised, knob-like section of some clam and mussel shells.
- *Water cycle*—is the biogeochemical cycle that moves and recycles water in various forms through the biosphere.
- *Wedge shaped*—means that a form is thick at one edge and tapered to a thin edge at the other.

1.3 LEVELS OF ORGANIZATION

Odum explains that "the best way to delimit modern ecology is to consider the concept of levels of organization."[12] Levels of organization can be simplified as follows:

Organs → Organism → Population → Communities → Ecosystem → Biosphere

In this relationship, organs form an organism; organisms of a particular species form a population; populations occupying a particular area form a community; communities, interacting with nonliving or abiotic factors, separate in a natural unit to create a stable system known as the *ecosystem* (the major ecological unit); and the part of the earth in which the ecosystem operates is known as the *biosphere*. Tomera points out that "every community is influenced by a particular set of abiotic factors."[13] The abiotic part of the ecosystem is represented by inorganic substances such as oxygen, carbon dioxide, several other inorganic substances, and some organic substances.

The physical and biological environment in which an organism lives is referred to as its *habitat*. For example, the habitat of two common aquatic insects, the "backswimmer" (*Notonecta*) and the "water boatman" (*Corixa*) is the littoral zone of ponds and lakes (shallow, vegetation-choked areas) (see Figure 1.1).[14]

Figure 1.1 *Notonecta* (left) and *Corixa* (right). (Source: Adapted from *Basic Ecology* by Eugene P. Odum, copyright © 1983 Saunders College Publishing, p. 402.)

[12]Odum, E. P., *Basic Ecology*. Philadelphia: Saunders College Publishing, p. 3, 1983.
[13]Tomera, A. N., *Understanding Basic Ecological Concepts*. Portland, ME: J. Weston Walch, Publisher, p. 41, 1989.
[14]Odum, E. P., *Basic Ecology*. Philadelphia: Saunders College Publishing, p. 402, 1983.

Within each level of organization of a particular habitat, each organism has a special role. The role the organism plays in the environment is referred to as its *niche*. A niche might be that the organism is food for some other organism or is a predator of other organisms. Odum refers to an organism's niche as its "profession."[15] Each organism has a job or role to fulfill in its environment. Although two different species might occupy the same habitat, "niche separation based on food habits" differentiates between two species.[16] Such niche separation can be seen by comparing the niches of the water backswimmer and the water boatman. The backswimmer is an active predator, while the water boatman feeds largely on decaying vegetation.[17]

✓ *Note:* In order for an ecosystem to exist, a dynamic balance must be maintained among all biotic and abiotic factors—a concept known as *homeostasis*.

1.4 ECOSYSTEM

Ecosystem is a term introduced by Tansley to denote an area that includes all organisms therein and their physical environment. The ecosystem is the major ecological unit in nature. "There is a constant interchange of the most various kinds within each system, not only between the organisms but between the organic and the inorganic."[18] Living organisms and their nonliving environment are inseparably interrelated and interact upon each other to create a self-regulating and self-maintaining system. To create a self-regulating and self-maintaining system, ecosystems are homeostatic, i.e., they resist any change through natural controls. These natural controls are important in ecology. People, through their complex activities, tend to disrupt natural controls.

As stated earlier, the ecosystem encompasses both the living and nonliving factors in a particular environment. The living or biotic part of the ecosystem is formed by two components: *autotrophic* and *heterotrophic*. The autotrophic (self-nourishing) component does not require food from its environment but can manufacture food from inorganic substances. For example, some autotrophic components (plants) manufacture needed energy through photosynthesis. Heterotrophic components, on the other hand, depend upon autotrophic components for food.[19]

The nonliving or abiotic part of the ecosystem is formed by three components: inorganic substances, organic compounds (link biotic and abiotic parts),

---

[15]Odum, E. P., *Ecology: The Link Between the Natural and the Social Sciences.* New York: Holt, Rinehart and Winston, p. 46, 1975.
[16]Odum, E. P., *Basic Ecology.* Philadelphia: Saunders College Publishing, p. 402, 1983.
[17]McCafferty, P. W., *Aquatic Entomology.* Boston: Jones and Bartlett Publishers, Inc., p. 67, 1981.
[18]Tansley, A. G., "The use and abuse of vegetation concepts and terms." *Ecology,* 16. p. 299, 1935.
[19]Porteous, A., *Dictionary of Environmental Science and Technology.* New York: John Wiley & Sons, Inc., p. 34, 1992.

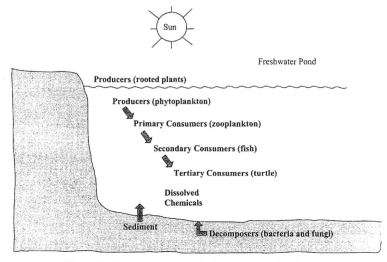

Figure 1.2 Major components of a freshwater pond ecosystem.

and climate regime. Figure 1.2 is a simplified diagram showing a few of the living and nonliving components of an ecosystem found in a freshwater pond.

An ecosystem is a cyclic mechanism in which biotic and abiotic materials are constantly exchanged through *biogeochemical cycles*. Biogeochemical cycles are defined as follows: "bio" refers to living organisms and "geo" refers to water, air, rocks, or solids. "Chemical" is concerned with the chemical composition of the earth. Biogeochemical cycles are driven by energy, directly or indirectly, from the sun. They will be discussed later in Chapter 3, "Biogeochemical Cycles."

Figure 1.2 depicts an ecosystem where biotic and abiotic materials are constantly exchanged. Producers construct organic substances through photosynthesis and chemosynthesis. Consumers and decomposers use organic matter as their food and convert it into abiotic components. That is, they dissipate energy fixed by producers through food chains. The abiotic part of the pond in Figure 1.2 is formed of inorganic and organic compounds that are dissolved and that are found in sediments, such as carbon, oxygen, nitrogen, sulfur, calcium, hydrogen, and humic acids. The biotic part is represented by producers such as rooted plants and phytoplanktons. Fish, crustaceans, and insect larvae make up the consumers. Detrivores, which feed on organic detritus, are represented by mayfly nymphs. Decomposers make up the final abiotic part. They include aquatic bacteria and fungi, which are distributed throughout the pond.

As stated earlier, an ecosystem is a cyclic mechanism. From a functional viewpoint, an ecosystem can be analyzed in terms of several factors. The factors important in this study include biogeochemical cycles, energy, and food chains.

1.5 SUMMARY OF KEY TERMS

- *Ecology*—is the study of the interrelationship of an organism or a group of organisms and their environment.
- *Ecosystem*—is the major ecological unit that has structure and function.
- *Environment*—is everything that is important to an organism in its surroundings. The *abiotic factor* is composed of sunlight, soil, mineral elements, temperature, moisture, and topography. The *biotic community* is the natural combination of organisms (plants and animals) that share the same area, are mutually sustaining, interdependent, and constantly fixing, utilizing, and dissipating energy.
- *Autotrophs*—(green plants) fix energy of the sun and manufacture food from simple, inorganic substances.
- *Heterotrophs*—(animals) use food stored by the autotroph, rearrange it, and finally decompose complex materials into simple inorganic compounds. Heterotrophs may be carnivorous (meat-eaters), herbivorous (plant-eaters), or omnivorous (plant- and meat-eaters).
- *Biogeochemical cycles*—are cyclic mechanisms in all ecosystems by which biotic and abiotic materials are constantly exchanged.
- *Consumers and decomposers*—dissipate energy fixed by the producers through food chains or webs. The available energy decreases by 80–90% during transfer from one trophic level to another.

1.6 CHAPTER REVIEW QUESTIONS

✓ *Note:* Answers to chapter review questions are found in Appendix A.

1.1 Another word for environmental interrelationship is _____.

1.2 Define autecology.

1.3 Define synecology.

1.4 What is point source?

1.5 Define pollution.

1.6 List eight abiotic factors found in a typical ecosystem:

  (1)

  (2)

  (3)

  (4)

  (5)

  (6)

  (7)

  (8)

Matching exercise: Match the definition listed in Part A with the terms listed in Part B by placing the correct letters in the blanks.

Part A:

  (1) The liquid wastes from a community are _____.
  (2) The study of organisms at home is _____.
  (3) The nonliving part of the environment is the _____.

(4) Less than 5 ppm of DO results in _____.
(5) The community and nonliving environment function together as an _____.
(6) The role that an organism plays in its natural ecosystem is its _____.
(7) Compatible association between dissimilar organisms is _____.
(8) The living part of the environment is the _____.
(9) The natural aging of a lake is _____.
(10) The diversity of species in an ecosystem is indicated by the _____.
(11) All of the populations occupying a given area comprise a _____.
(12) The mixing of air and water is called _____.
(13) Agricultural runoff is an example of _____.
(14) An example of a natural process is a _____.
(15) A group of organisms of a single species is called a _____.
(16) The energy of the sun is fixed by an _____.
(17) Fish, crustaceans, and insect larvae are _____.
(18) Modern ecology is delimited by _____.
(19) When something is divided into similar, repeated sections it is _____.
(20) Nourishment is provided by _____.
(21) A local habitat is called a _____.
(22) A juvenile form of an insect is a _____.
(23) An underground stream flow is called _____.
(24) Chlorophyll is lacking in _____.
(25) Matter floating along with stream flow is _____.
(26) Rapid flow over a shallow area is called _____.

Part B:

a. aeration
b. fungi
c. sewage
d. non-point pollution
e. riffle
f. ecology
g. drift
h. water cycle
i. ecosystem
j. segmented
k. eutrophication
l. hyporeal
m. nutrients
n. abiotic factor
o. population
p. microhabitat
q. fishkill
r. biotic index
s. larva
t. niche
u. autotroph
v. community
w. levels of organization
x. biotic factor
y. consumers
z. symbiosis

INTRODUCTION

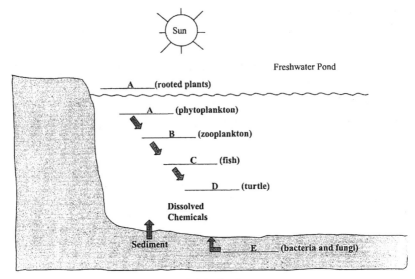

Label the Pond.

Label the Pond Exercise: Identify the lettered components in the drawing. Insert the correct descriptions from the list to correspond with the correct number.

List:

    secondary consumers
    producers
    tertiary consumers
    decomposers
    producers

A. _____

B. _____

C. _____

D. _____

E. _____

CHAPTER 2

Stream Genesis and Structure

Early in the spring on a snow- and ice-covered high alpine meadow, the water cycle continues. The cycle's main component, water, has been held in reserve, literally frozen for the winter months, but with spring, the sun is higher, more direct, and of longer duration, and the frozen masses of water begin to melt. This snowmelt makes its way down the mountain, forming pools. Waters from higher elevations flow into the pools, and the overflow caused by this continues the flow of the melt water. The waters become progressively discolored as they pass over the terrain, stained brown-black with humic acid and filled with suspended sediments.

The waters divide and flow in different directions, over different landscapes. Small streams divert and flow into open country. Different soils work to retain or speed the waters, and in some places, the waters spread into shallow swamps, bogs, marshes, fens, or mires. Other streams pause to fill depressions in the land and to form lakes, short-term resting places in the water cycle. The water is eventually evaporated or seeps into groundwater. Other portions of the water mass stay with the main flow, and the speed of flow changes to form a river. As it changes speed and slows, the river bottom changes from rock and stone to silt and clay. Plants begin to grow, stems thicken, and leaves broaden. The river now provides the nutrients needed to sustain life. Eventually, the river drains into the sea.[20]

2.1 INTRODUCTION

THE main point to be gained from the chapter opening is that the physical processes involved in the formation of a stream are important to the ecology of the stream. Stream channel and flow characteristics directly influence the functioning of the stream's ecosystem and the biota found therein. Thus, in this chapter, we look at the pathways of water flow contributing to stream flow; namely, we discuss precipitation inputs as they contribute to flow. Stream flow dis-

---

[20]Spellman, F. R. and Whiting, N., *Environmental Science & Technology*. Rockville, MD: Government Institutes, pp. 265–267, 1999.

charge, transport of material, characteristics of stream channels, stream profile, sinuosity, the floodplain, pool-riffle sequences, and depositional features—all of which directly or indirectly impact the ecology of the stream—are also discussed.

2.2 WATER FLOW IN A STREAM

In this text, we are primarily concerned with the surface water route taken by surface water runoff. Surface runoff is dependent on various factors. For example, climate, vegetation, topography, geology, soil characteristics, and land-use determine how much surface runoff occurs compared with other pathways.

The primary source (input) of water to total surface runoff, of course, is precipitation. This is the case even though a substantial portion of all precipitation input returns directly to the atmosphere by evapotranspiration. *Evapotranspiration* is a combination process whereby water in plant tissue and in soil evaporates and transpires to water vapor in the atmosphere.

A substantial portion of precipitation input returns directly to the atmosphere by evapotranspiration. When precipitation occurs, some rainwater is intercepted by vegetation where it evaporates, never reaching the ground or being absorbed by plants. A large portion of the rainwater that reaches the surface, on ground, in lakes, and in streams, also evaporates directly back to the atmosphere. Although plants display a special adaptation to minimize transpiration, plants still lose water to the atmosphere during the exchange of gases necessary for photosynthesis. Notwithstanding the large percentage of precipitation that evaporates, rain- or melt-water that reaches the ground surface follows several pathways in reaching a stream channel or groundwater.

Soil can absorb rainfall to its *infiltration capacity* (i.e., to its maximum rate). During a rain event, this capacity decreases. Any rainfall in excess of infiltration capacity accumulates on the surface. When this surface water exceeds the depression storage capacity of the surface, it moves as an irregular sheet of overland flow. In arid areas, overland flow is likely due to the low permeability of the soil. Overland flow is also likely when the surface is frozen and/or when human activities have rendered the land surface less permeable. In humid areas, where infiltration capacities are high, overland flow is rare.

In rain events where the infiltration capacity of the soil is not exceeded, rain penetrates the soil and eventually reaches the groundwater, from which it discharges to the stream slowly, and over a long period of time. This phenomenon helps to explain why stream flow through a dry weather region remains constant; the flow is continuously augmented by groundwater. This type of stream is known as a *perennial* stream, as opposed to an *intermittent* one, because the flow continues during periods of no rainfall.

Streams that course their way through humid regions are fed water via the

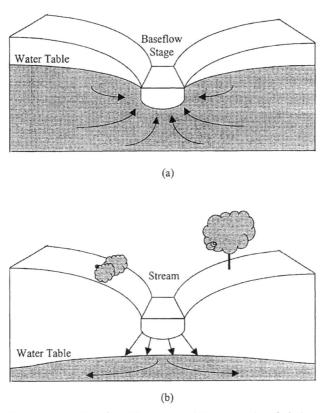

Figure 2.1 (a) Cross section of a gaining stream and (b) cross section of a losing stream.

water table, which slopes toward the stream channel. Discharge from the water table into the stream accounts for flow during periods without precipitation and also explains why this flow increases, even without tributary input, as one proceeds downstream. Such streams are called *gaining* or *effluent,* as opposed to *losing* or *influent* streams that lose water into the ground (see Figure 2.1). It is interesting to note that the same stream can shift between gaining and losing conditions along its course because of changes in underlying strata and local climate.

2.3 STREAM WATER DISCHARGE

The current velocity (speed) of water (driven by gravitational energy) in a channel varies considerably within a stream's cross section owing to friction with the bottom and sides, to sediment, to the atmosphere, and to sinuosity (bending or curving) and obstructions. The highest velocities are generally found at or near the surface and near the center of the channel, where there is the least amount of friction. In deeper streams, current velocity is greatest just below the surface due to the friction with the atmosphere; in shallower streams, current velocity is greatest at the surface due to friction with the bed. Velocity decreases as a function of depth, approaching zero at the substrate surface.

2.4 TRANSPORT OF MATERIAL

Water flowing in a channel may exhibit *laminar flow* (parallel layers of water shear over one another vertically) or *turbulent flow* (complex mixing). In streams, laminar flow is uncommon, except at boundaries where flow is very low and in groundwater. Thus, the flow in streams is generally turbulent. Turbulence exerts a shearing force that causes particles to move along the stream-bed by pushing, rolling, and skipping (referred to as *bed load*). This same shear causes turbulent eddies that entrain particles in suspension (called the *suspended load*—particles under 0.06 mm).

Entrainment is the incorporation of particles when stream velocity exceeds the *entraining velocity* for a particular particle size.

✓ *Note:* Entrainment is a natural extension of erosion and is vital to the movement of stationary particles in changing flow conditions. Remember, all sediments ultimately derive from erosion of basin slopes, but the immediate supply usually derives from the stream channel and banks, while the bed load comes from the streambed itself and is replaced by erosion of bank regions.

The entrained particles in suspension (suspended load) also include fine sediment, primarily clays, silts, and fine sands, that require only low velocities and minor turbulence to remain in suspension. These are referred to as *wash load* (under 0.002 mm), because this load is "washed" into the stream from banks and upland areas.[21]

Thus, the suspended load includes the wash load and coarser materials (at lower flows). Together, the suspended load and bed load constitute the *solid load*. It is important to note that in bedrock streams, the bed load will be a lower fraction than in alluvial streams where channels are composed of easily transported material.

A substantial amount of material is also transported as the *dissolved load*. Solutes are generally derived from chemical weathering of bedrock and soils,

---

[21]Gordon, N. D., McMahon, T. A., and Finlayson, B. L., *Stream Hydrology: An Introduction for Ecologists.* Chichester, UK: Wiley, p. 4, 1992.

and their contribution is greatest in subsurface flows and in regions of limestone geology.

The relative amount of material transported as solute rather than solid load depends on basin characteristics, lithology (i.e., the physical character of rock), and hydrologic pathways. In areas of very high runoff, the contribution of solutes approaches or exceeds sediment load, whereas in dry regions, sediments make up as much as 90% of the total load.

Deposition occurs when *stream competence* (i.e., the largest particle that can be moved as bed load, and the critical erosion—competent—velocity is the lowest velocity at which a particle resting on the streambed will move) falls below a given velocity. Simply stated: the size of the particle that can be eroded and transported is a function of current velocity.

Sand particles are the most easily eroded. The greater the mass of larger particles (e.g., coarse gravel), the higher the initial current velocities must be for movement. However, smaller particles (silts and clays) require even greater initial velocities because of their cohesiveness and because they present smaller, streamlined surfaces to the flow. Once in transport, particles will continue in motion at somewhat slower velocities than initially required to initiate movement, and they will settle at still lower velocities.

Particle movement is determined by size, flow conditions, and mode of entrainment. Particles over 0.02 mm (medium-coarse sand size) tend to move by rolling or sliding along the channel bed as *traction load*. When sand particles fall out of the flow, they move by *saltation* or repeated bouncing. Particles under 0.06 mm (silt) move as *suspended load*, and particles under 0.002 (clay), indefinitely, as *wash load*. A considerable amount of particle sorting takes place because of the different styles of particle flow in different sections of the stream.[22]

Unless the supply of sediments becomes depleted, the concentration and amount of transported solids increase. However, discharge is usually too low throughout most of the year to scrape or scour, shape channels, or move significant quantities of sediment in all but sand-bed streams, which can experience change more rapidly. During extreme events, the greatest scour occurs, and the amount of material removed increases dramatically.

Sediment inflow into streams can be increased and decreased as a result of human activities. For example, poor agricultural practices and deforestation greatly increase erosion.

Man-made structures such as dams and channel diversions can greatly reduce sediment inflow.

---

[22]Richards, K., *Rivers: Form and Processes in Alluvial Channels*. London: Methuen, p. 69, 1982; Likens, W. M., "Beyond the Shoreline: A Watershed Ecosystem Approach." *Vert. Int. Ver. Theor. Awg Liminol.*, 22, 1–22, 1984.

2.5 CHARACTERISTICS OF STREAM CHANNELS

Flowing waters (rivers and streams) determine their own channels, and these channels exhibit relationships attesting to the operation of physical laws—laws that are not, as of yet, fully understood. The development of stream channels and entire drainage networks, and the existence of various regular patterns in the shape of channels, indicate that streams are in a state of dynamic equilibrium between erosion (sediment loading) and deposition (sediment deposit) and are governed by common hydraulic processes. However, because channel geometry is four dimensional with a long profile, cross section, depth and slope profile, and because these mutually adjust over a time scale as short as years and as long as centuries or more, cause and effect relationships are difficult to establish. Other variables that are presumed to interact as the stream achieves its graded state include width and depth, velocity, size of sediment load, bed roughness, and the degree of braiding (sinuosity).

2.6 STREAM PROFILES

Mainly because of gravity, most streams exhibit a downstream decrease in gradient along their length. Beginning at the headwaters, the steep gradient becomes less as one proceeds downstream, resulting in a concave longitudinal profile. Though diverse geography provides for almost unlimited variation, a lengthy stream that originates in a mountainous area (such as the one described in the chapter opening) typically comes into existence as a series of springs and rivulets; these coalesce into a fast-flowing, turbulent mountain stream, and the addition of tributaries results in a large and smoothly flowing river that winds through the lowlands to the sea.

When studying a stream system of any length, it becomes readily apparent that it is a body of flowing water that varies considerably from place to place along its length. For example, a common variable is that whenever discharge increases, corresponding changes in the stream's width, depth, and velocity can be readily seen. In addition to physical changes that occur from location to location along a stream's course, there are biological variables that correlate with stream size and distance downstream. The most apparent and striking changes are in steepness of slope and in the transition from a shallow stream with large boulders and a stony substrate to a deep stream with a sandy substrate.

The particle size of bed material at various locations is also variable along the stream's course. The particle size usually shifts from an abundance of coarser material upstream to mainly finer material downstream.

2.7 SINUOSITY

Unless forced by man in the form of heavily regulated and channelized streams, straight channels are uncommon. Stream flow creates distinctive landforms composed of straight (usually in appearance only), meandering, and braided channels, channel networks, and flood plains. Flowing water will follow a sinuous course. The most commonly used measure is the *sinuosity index* (SI). Sinuosity equals one in straight channels and more than one in sinuous channels.

$$SI = \frac{\text{channel distance}}{\text{down valley distance}} \quad (2.1)$$

Meandering is the natural tendency for alluvial channels and is usually defined as an arbitrarily extreme level of sinuosity, typically an SI greater than 1.5. Many variables affect the degree of sinuosity, however, and SI values range from near unity in simple, well-defined channels to four in highly meandering channels.[23]

It is interesting to note that even in many natural channel sections of a stream course that appear straight, meandering occurs in the line of maximum water or channel depth (known as the *thalweg*). A stream renews itself by meandering. Streams wash plants and soil from the land into their waters, and these serve as nutrients for the plants in the rivers. If rivers aren't allowed to meander, if they are channelized, the amount of life they can support will gradually decrease. That means less fish, ultimately, and fewer bald eagles, herons, and other fishing birds.[24]

Meander flow follows a predictable pattern and causes regular regions of erosion and deposition (see Figure 2.2). The streamlines of maximum velocity and the deepest part of the channel lie close to the outer side of each bend and cross over near the point of inflection between the banks (see Figure 2.2). A huge elevation of water at the outside of a bend causes a helical flow of water toward the opposite bank. In addition, a separation of surface flow causes a back eddy. The result is zones of erosion and deposition, and explains why point bars develop in a downstream direction in depositional zones.[25]

✓ *Note:* Meandering channels can be highly convoluted or merely sinuous but maintain a single thread in curves having definite geometric shape. Straight channels are sinuous but apparently random in occurrence of bends. Braided channels are those with multiple streams separated by bars and islands.[26]

---

[23]Gordon, N. D., McMahon, T. A., and Finlayson, B. L., *Stream Hydrology: An Introduction for Ecologists.* Chichester, UK: Wiley, p. 49, 1992.
[24]Cave, C., *How a River Flows.* http://home.netcom.com/~cristi, cristi@ix.netcom.com, p. 3, 2000.
[25]Morisawa, M., *Streams: Their Dynamics and Morphology.* New York: McGraw-Hill, p. 66, 1968.
[26]Leopold, L. B., *A View of the River.* Cambridge, MA: Harvard University Press, p. 56, 1994.

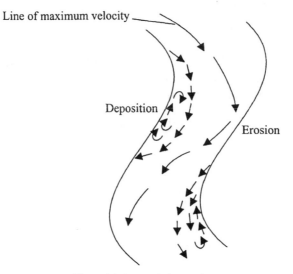

Figure 2.2 A meandering reach.

(a)

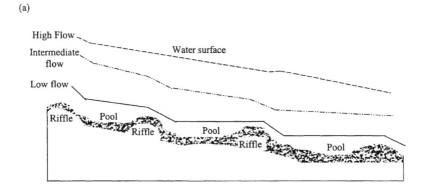

(b)

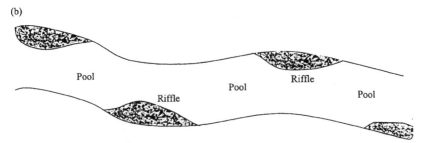

Figure 2.3 (a) Longitudinal profile of a riffle-pool sequence and (b) plain view of a riffle-pool sequence.

2.8 BARS, RIFFLES, AND POOLS

Implicit in the morphology and formation of meanders are bars, riffles, and pools. Bars develop by deposition in slower, less competent flow on either side of the sinuous mainstream. Onward moving water, depleted of bed load, regains competence and shears a pool in the meander, reloading the stream for the next bar. Alternating bars migrate to form riffles (see Figure 2.3).

As stream flow continues along its course, a pool-riffle sequence is formed. Basically, the riffle is a mound or hillock, and the pool is a depression.

2.9 THE FLOODPLAIN

Stream channels influence the shape of the valley floor through which they course. This self-formed, self-adjusted flat area near the stream is the floodplain, which loosely describes the valley floor prone to periodic inundation during over-bank discharges. What is not commonly known is that valley flooding is a regular and natural behavior of the stream. Many people learn about this natural phenomenon whenever their towns, streets, and homes become inundated by a river or stream that is following its "natural" periodic cycle.

✓ *Note:* Floodplain rivers are found where regular floods form lateral plains outside the normal channel which seasonally become inundated, either as a consequence of greatly increased rainfall or snow melt.[27]

2.10 SUMMARY OF KEY TERMS

- *Evapotranspiration (plant water loss)*—describes the process whereby plants lose water to the atmosphere during the exchange of gases necessary for photosynthesis. Water loss by evapotranspiration constitutes a major flux back to the atmosphere.
- *Infiltration capacity*—is the maximum rate soil can absorb rainfall.
- *Perennial stream*—is a type of stream in which flow continues during periods of no rainfall.
- *Gaining stream*—is typical of humid regions, where groundwater recharges the stream.
- *Losing stream*—is typical of arid regions, where streams can recharge groundwater.
- *Laminar flow*—occurs in a stream where parallel layers of water shear over one another vertically.

---

[27]Giller, P. S. and Jalmqvist, B., *The Biology of Streams and Rivers*. Oxford, UK: Oxford University Press, p. 26, 1998.

24 STREAM GENESIS AND STRUCTURE

- *Turbulent flow*—occurs in a stream where complex mixing is the result.
- *Meandering*—is a stream condition whereby flow follows a winding and turning course.
- *Thalweg*—is a line of maximum water of channel depth in a stream.
- *Riffles*—refers to shallow, high-velocity flow over mixed gravel-cobble (bar-like) substrate.
- *Sinuosity*—is the bending or curving shape of a stream course.

2.11 CHAPTER REVIEW QUESTIONS

2.1 The particle size usually _____ from an abundance of _____ material upstream to mainly _____ material in downstream areas.

2.2 The primary source of water to total surface runoff is _____.

2.3 Define evapotranspiration.

2.4 Soil's _____ is the maximum amount of rainfall it can absorb.

2.5 The type of stream in which flow continues during periods of no rainfall is a _____.

2.6 In stream flow, the highest velocities are found where?

2.7 Define entrainment.

2.8 From where are stream sediments ultimately derived?

2.9 When sand particles fall out of the flow, they move by _____.

2.10 _____ develop by deposition in slower, less competent flow on either side of the sinuous mainstream.

2.11 The line of maximum water of channel depth in a stream is known as _____.

2.12 Define sinuosity.

CHAPTER 3

Biogeochemical Cycles

Water is Earth's "proud setter up and puller down of kings."[28]

3.1 NUTRIENT CYCLES

STREAMS and rivers are complex ecosystems that take part in the physical and chemical cycles (biogeochemical cycles) that shape our planet and allow life to exist.[29] A *biogeochemical cycle* is composed of *bioelements* (chemical elements that cycle through living organisms), and it occurs when there is interaction between the biological and physical exchanges of bioelements.

✓ *Note:* Contrary to an incorrect assumption, energy does not cycle through an ecosystem—chemicals do. The inorganic nutrients cycle through more than the organisms; however, they also enter the oceans, atmosphere, and even rocks. Because these chemicals cycle through both the biological and the geological worlds, we call the overall cycles biogeochemical cycles.

Each chemical has a unique cycle, but all cycles have some things in common. Reservoirs are those parts of the cycle where the chemical is held in large quantities for long periods of times (e.g., the oceans for water and rocks for phosphorous). In exchange pools, on the other hand, the chemical is held for only a short time (e.g., the atmosphere; a cloud). The length of time a chemical is held in an exchange pool or a reservoir is termed its residence time. The biotic community includes all living organisms. This community may serve as an exchange pool (although for some chemicals like carbon that can be bound in certain tree species for a thousand years, it may seem more like a reservoir), and it

---

[28]Shakespeare's description of Richard—the Kingmaker during the Wars of the Roses.
[29]Cave, C., *Stream Ecology*. http://home.netcom.com/~cristi, p. 1, 1998.

also may serve to move chemicals (bioelements) from one stage of the cycle to another. For instance, the trees of the tropical rain forest bring water up from the forest floor to be transpired into the atmosphere. Likewise, coral organisms take carbon from the water and turn it into limestone rock. The energy for most of the transportation of chemicals is provided either by the sun or by the heat released from the mantle and core of the earth.

In the case of chemical elements that cycle through living things, the following can be stated:[30]

- All bioelements reside in compartments or defined spaces in nature.
- A compartment contains a certain quantity, or pool, of bioelements.
- Compartments exchange bioelements. The rate of movement of bioelements between two compartments is called the flux rate.
- The average length of time a bioelement remains in a compartment is called the mean residence time (MRT).
- The flux rate and pools of bioelements together define the nutrient cycle in an ecosystem.
- Ecosystems are not isolated from one another, and bioelements come into an ecosystem through meteorological, geological, or biological transport mechanisms:
  —meteorological (e.g., deposition in rain and snow, atmospheric gases)
  —geological (e.g., surface and subsurface drainage)
  —biological (e.g., movement of organisms between ecosystems)

As a result, biogeochemical cycles can be local or global.

Smith categorizes biogeochemical cycles into two types, the *gaseous* and the *sedimentary*. Gaseous cycles include carbon and nitrogen cycles. The main pool (or sink) of nutrients in the gaseous cycle is the atmosphere and the ocean. The sedimentary cycles include sulfur and phosphorous cycles. The main sink for sedimentary cycles is soil and rocks of the earth's crust.[31]

Between 20 to 40 of the earth's 92 naturally occurring elements are ingredients that make up living organisms. The chemical elements carbon, hydrogen, oxygen, nitrogen, and phosphorus are critical in maintaining life. Odum points out that of the elements needed by living organisms to survive, oxygen, hydrogen, carbon, and nitrogen are needed in larger quantities than some of the other elements.[32] These elements exhibit definite biogeochemical cycles, which will be discussed in detail later.

The elements needed to sustain life are products of the global environment that consists of three main subdivisions (see Figure 3.1):

---

[30]From *Biogeochemical Cycles II: The Nitrogen and Phosphorus Cycles.* Yahoo.com internet access, pp. 1–2, 2000.
[31]Smith, R. L., *Ecology and Field Biology.* New York: Harper & Row, p. 49, 1974.
[32]Odum, E. P., *Fundamentals of Ecology.* Philadelphia: Saunders College Publishing, p. 30, 1971.

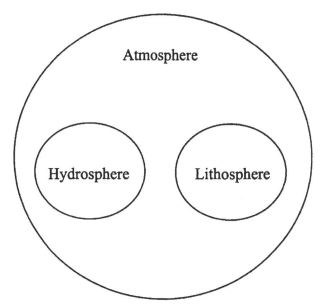

Figure 3.1 The global environment.

(1) *Hydrosphere*—includes all components formed of water bodies on the earth's surface.
(2) *Lithosphere*—comprises the solid components on the earth's surface such as rocks.
(3) *Atmosphere*—is the gaseous mantle that envelops the hydrosphere and the lithosphere.

To survive, organisms require inorganic metabolites from all three parts of the biosphere. For example, the hydrosphere supplies water as the exclusive source of needed hydrogen. Essential elements such as calcium, sulfur, and phosphorus are provided by the lithosphere. Finally, oxygen, nitrogen, and carbon dioxide are provided by the atmosphere.

Within the biogeochemical cycles, all the essential elements circulate from the environment to organisms and back to the environment. Because of the critical importance of elements in sustaining life, it may be easily understood why biogeochemical cycles are readily and realistically labeled *nutrient cycles.*

Through these biogeochemical or nutrient cycles, nature processes and reprocesses the critical life-sustaining elements in definite inorganic-organic cycles. In some cycles, such as carbon, there is no loss of material for long periods of time. One point to keep in mind is that energy (to be explained later) flows "through" an ecosystem, but nutrients are cycled and recycled.

Humans need most of these recycled elements to survive. Because of this, we have speeded up the movement of many materials so that the cycles tend to become imperfect, or what Odum calls *acyclic*. Odum goes on to explain that our environmental impact on phosphorus demonstrates one example of a somewhat imperfect cycle.

> We mine and process phosphate rock with such careless abandon that severe local pollution results near mines and phosphate mills. Then, with equally acute myopia we increase the input of phosphate fertilizers in agricultural systems without controlling in any way the inevitable increase in run-off output that severely stresses our waterways and reduces water quality through eutrophication.[33]

As related above, in agricultural ecosystems, we often supply necessary nutrients in the form of fertilizer to increase plant growth and yield. In natural ecosystems, however, these nutrients are recycled naturally through each trophic level. For example, the elemental forms are taken up by plants. The consumers ingest these elements in the form of organic plant material. Eventually, the nutrients are degraded to the inorganic form again. The following pages present and discuss the nutrient cycles for carbon, nitrogen, phosphorus, and sulfur.

3.2 CARBON CYCLE

Carbon, which is an essential ingredient of all living things, is the basic building block of the large organic molecules necessary for life. Carbon is cycled into food chains from the atmosphere, as shown in Figure 3.2.

The carbon cycle (see Figure 3.2) is based on carbon dioxide, which makes up only a small percentage of the atmosphere. From Figure 3.2, it can be seen that green plants obtain carbon dioxide (CO_2) from the air and, through photosynthesis, described by Asimov as the "most important chemical process on Earth," it produces the food and oxygen on which all organisms live.[34] Part of the carbon produced remains in living matter, and the other part is released as CO_2 in cellular respiration. Miller points out that the carbon dioxide released by cellular respiration in all living organisms is returned to the atmosphere.[35]

✓ *Note:* About a tenth of the estimated 700 billion tons of carbon dioxide in the atmosphere is fixed annually by photosynthetic plants. A further trillion tons are dissolved in the ocean, more than half in the photosynthetic layer.

Some carbon is contained in buried dead animal and plant materials. Much of these buried plant and animal materials were transformed into fossil fuels. Fossil fuels (coal, oil, and natural gas) contain large amounts of carbon. When

---

[33]Odum, E. P., *Fundamentals of Ecology*. Philadelphia: Saunders College Publishing, p. 87, 1971.
[34]Asimov, I., *How Did We Find Out About Photosynthesis?* New York: Walker & Company, p. 20, 1989.
[35]Miller, G. T., *Environmental Science: An Introduction*. Belmont, CA: Wadsworth, p. 43, 1988.

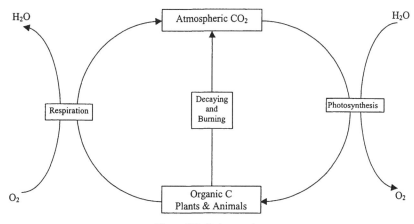

Figure 3.2 Carbon cycle.

fossil fuels are burned, stored carbon combines with oxygen in the air to form carbon dioxide, which enters the atmosphere.[36]

In the atmosphere, carbon dioxide acts as a beneficial heat screen as it does not allow the heat generated by earth's radiant energy to be emitted into space. This balance is important. The problem is that as more carbon dioxide from burning is released into the atmosphere, balance can and is being altered. Odum warns that recent increase in consumption of fossil fuels "coupled with the decrease in the 'removal capacity' of the green belt is beginning to exceed the delicate balance."[37] Increased releases of carbon dioxide into the atmosphere tend to increase the possibility of global warming. The consequences of global warming "would be catastrophic . . . and the resulting climatic change would be irreversible."[38]

3.3 NITROGEN CYCLE

Nitrogen is important to all life. Nitrogen in the atmosphere or in the soil can go through many complex chemical and biological changes, be combined into living and nonliving material, and return to the soil or air in a continuing cycle. This is called the nitrogen cycle.[39]

The atmosphere contains 78% by volume of nitrogen. Moreover, as stated

---

[36]Moran, J. M., Morgan, M. D., and Wiersma, J. H., *Introduction to Environmental Science.* New York: W.H. Freeman and Company, p. 67, 1986.
[37]Odum, E. P., *Basic Ecology.* Philadelphia: Saunders College Publishing, p. 202, 1983.
[38]Abrahamson, D. E. (ed.). *The Challenge of Global Warming.* Washington, DC: Island Press, p. 4, 1988.
[39]Killpack, S. C. and Buchholz, D., *Nitrogen in the Environment: Nitrogen.* Missouri: University of Missouri-Columbia, p. 1, 1993.

previously, nitrogen is an essential element for all living matter and constitutes 1–3% dry weight of cells, yet nitrogen is not a common element on earth. Although it is an essential ingredient for plant growth, it is chemically very inactive, and before it can be incorporated by the vast majority of the biomass, it must be fixed.[40]

Price describes the nitrogen cycle as an example "of a largely complete chemical cycle in ecosystems with little leaching out of the system."[41] From the water/wastewater specialist's point of view, nitrogen and phosphorous are both commonly considered limiting factors for productivity. Of the two, nitrogen is harder to control but is found in smaller quantities in wastewater.

As stated earlier, nitrogen gas makes up about 78% of the volume of the earth's atmosphere. As such, it is useless to most plants and animals. Fortunately, nitrogen gas is converted into compounds containing nitrate ions, which are taken up by plant roots as part of the nitrogen cycle, shown in simplified form in Figure 3.3.

Aerial nitrogen is converted into nitrates mainly by microorganisms, bacteria, and blue-green algae. Lightning also converts some aerial nitrogen gas into forms that return to the earth as nitrate ions in rainfall and other types of precipitation. From Figure 3.3, it can be seen that ammonia plays a major role in the nitrogen cycle. Excretion by animals and anaerobic decomposition of dead organic matter by bacteria produce ammonia. Ammonia, in turn, is converted by nitrification bacteria into nitrites and then into nitrates. This process is known as *nitrification*. Nitrification bacteria are aerobic. Bacteria that convert ammonia into nitrites are known as nitrite bacteria (*Nitrosococcus* and *Nitrosomonas*); they convert nitrites into nitrates and nitrate bacteria (*Nitrobacter*). In wastewater treatment, ammonia is produced in the sludge digester and nitrates are produced in the aerobic sewage treatment process.

In *Wastewater Engineering*, several pages are devoted to describing the nitrogen cycle and its impact on the wastewater treatment process. They point out that nitrogen is found in wastewater in the form of urea. During wastewater treatment, the urea is transformed into ammonia nitrogen. Because ammonia exerts a BOD and chlorine demand, high quantities of ammonia in wastewater effluents are undesirable. The process of nitrification is utilized to convert ammonia to nitrates. Nitrification is a biological process that involves the addition of oxygen to the wastewater. If further treatment is necessary, another biological process called denitrification is used.[42] In this process, nitrate is converted into nitrogen gas, which is lost to the atmosphere, as can be seen in Figure 3.3.

When attempting to address the important and complex factors that make up

---

[40]Porteous, A., *Dictionary of Environmental Science and Technology*. New York: John Wiley & Sons, Inc., p. 83, 1992.
[41]Price, P. W., *Insect Ecology*. New York: John Wiley & Sons, Inc., p. 11, 1984.
[42]Metcalf & Eddy, Inc., *Wastewater Engineering: Treatment, Disposal, Reuse*. 3rd ed. New York: McGraw-Hill, pp. 85–87, 1991.

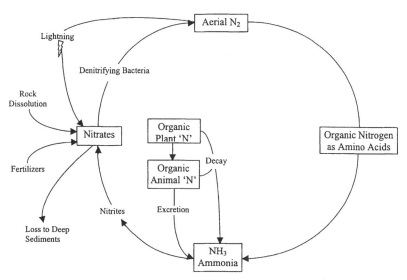

Figure 3.3 Nitrogen cycle.

the topic of stream ecology and self-purification, it is important to understand the impact that the nitrogen cycle can have on effluent that is dumped (outfalled) into receiving streams. At the same time, one should remember that the nitrogen cycle that occurs in the wastewater stream is not the source of the nitrogen contamination of surface water bodies. As a case in point, Price uses the example of large inputs of nitrogen fertilizer from agricultural systems, which "may result in considerable leaching and unidirection flow of nitrogen into aquatic systems which become polluted with excessive nitrogen . . . ."[43]

✓ *Note:* Nitrogen becomes a concern to stream ecology (quality) when nitrogen in the soil is converted to nitrate (NO_3^-) form. Nitrate is very mobile and moves with water in the soil. The concern of nitrates and water quality is generally directed at groundwater. However, nitrates can also enter surface waters such as ponds, streams, and rivers. Nitrates in drinking water can lead to nitrate poisoning in infant humans and animals, causing serious health problems and even death. This occurs because of a bacteria commonly found in the intestinal tract of infants that can convert nitrate to high toxic nitrites (NO_2). Nitrite can replace oxygen in the bloodstream and result in oxygen starvation that causes a bluish discoloration of the infant ("blue baby" syndrome).[44]

---

[43]Price, P. W., *Insect Ecology*. New York: John Wiley & Sons, Inc., p. 11, 1984.
[44]Spellman, F. R., *The Science of Water: Concepts & Applications*. Lancaster, PA: Technomic Publishing Company, Inc., pp. 175–176, 1998.

3.4 PHOSPHORUS CYCLE

Phosphorus is another element that is common in the structure of living organisms. However, of all the elements recycled in the biosphere, phosphorus is the scarcest and, therefore, the one most limiting in any given ecological system. It is indispensable to life, being intimately involved in energy transfer and in the passage of genetic information in the DNA of all cells.

The ultimate source of phosphorus is rock, from which it is released by weathering, leaching, and mining. Phosphorus occurs as phosphate or other minerals formed in past geological ages. These massive deposits are gradually eroding to provide phosphorus to ecosystems. A large amount of eroded phosphorus ends up in deep sediments in the oceans and lesser amounts in shallow sediments. Part of the phosphorus comes to land when marine animals are brought out. Birds also play a role in the recovery of phosphorus. The great guano deposit, bird excreta, of the Peruvian coast is an example. Man has hastened the rate of loss of phosphorus through mining activities and the subsequent production of fertilizers. Even with the increase in human activities, however, there is no immediate cause for concern, because the known reserves of phosphate are quite large. Figure 3.4 shows the phosphorus cycle.

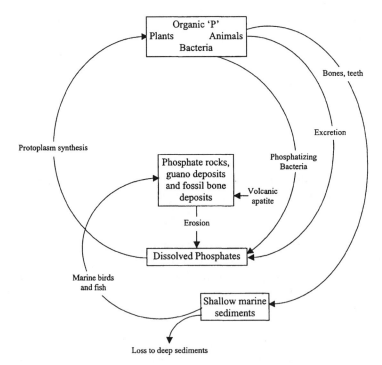

Figure 3.4 The phosphorus cycle.

Phosphorous has become very important in water quality studies, because it is often found to be a limiting factor (i.e., limiting plant nutrient). Control of phosphorus compounds that enter surface waters and contribute to growth of algal blooms is of much interest to stream ecologists. Phosphates, upon entering a stream, act as fertilizer, which promotes the growth of undesirable algae populations or algal blooms. As the organic matter decays, dissolved oxygen levels decrease, and fish and other aquatic species die.

While it is true that phosphorus discharged into streams is a contributing factor to stream pollution (and causes eutrophication), it is also true that phosphorus is not the lone factor. Odum warns against what he calls the one-factor control hypothesis, i.e., the one-problem/one-solution syndrome. He goes on to point out that environmentalists in the past have focused on one or two items, like phosphorous contamination, and "have failed to understand that the strategy for pollution control must involve reducing the input of all enriching and toxic materials."[45]

✓ *Note:* Because of its high reactivity, phosphorus exists in combined form with other elements. Microorganisms produce acids that form soluble phosphate from insoluble phosphorus compounds. The phosphates are utilized by algae and terrestrial green plants, which in turn pass into the bodies of animal consumers. Upon death and decay of organisms, phosphates are released for recycling.[46]

3.5 SULFUR CYCLE

Sulfur, like nitrogen and carbon, is characteristic of organic compounds. However, an important distinction between cycling of sulfur and cycling of nitrogen and carbon is that sulfur is "already fixed." That is, plenty of sulfate anions are available for living organisms to utilize. By contrast, the major biological reservoirs of nitrogen atoms (N_2) and carbon atoms (CO_2) are gases that must be pulled out of the atmosphere.

The sulfur cycle (see Figure 3.5) is both sedimentary and gaseous. Tchobanoglous and Schroeder note that "the principal forms of sulfur that are of special significance in water quality management are organic sulfur, hydrogen sulfide, elemental sulfur and sulfate."[47]

Bacteria play a major role in the conversion of sulfur from one form to another. In an anaerobic environment, bacteria break down organic matter pro-

---

[45]Odum, E. P., *Ecology: The Link Between the Natural and the Social Sciences.* New York: Holt, Rinehart and Winston, Inc., p. 110, 1975.
[46]*Phosphorus Cycle.* Britannica.com Inc., p. 1, 2000.
[47]Tchobanoglous, G. and Schroeder, E. D., *Water Quality.* Reading, MA: Addison-Wesley, p. 184, 1985.

BIOGEOCHEMICAL CYCLES

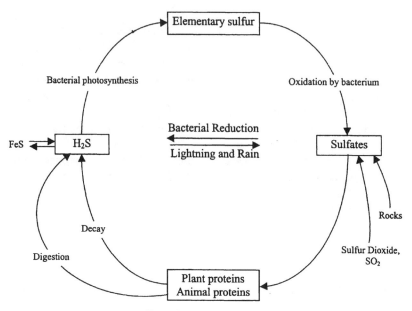

Figure 3.5 The sulfur cycle.

ducing hydrogen sulfide with its characteristic rotten egg odor. A bacteria called *Beggiatoa* converts hydrogen sulfide into elemental sulfur. An aerobic sulfur bacterium, *Thiobacillus thiooxidans*, converts sulfur into sulfates. Other sulfates are contributed by the dissolving of rocks and some sulfur dioxide. Sulfur is incorporated by plants into proteins. Some of these plants are then consumed by organisms. Sulfur from proteins is liberated by many heterotrophic anaerobic bacteria, as hydrogen sulfide.

3.6 SUMMARY OF KEY TERMS

- *Hydrosphere*—is the water covering the earth's surface of which 80% is salt, 19% is groundwater and, obviously, the other 1% is unsalted fresh surface water (rivers, lakes, streams, etc.).
- *Lithosphere*—is comprised of the solid components of the earth's surface such as rocks and weathered soil.
- *Atmosphere*—is the gaseous mantle enveloping the hydrosphere and lithosphere, which is 78% nitrogen by volume.
- *Organisms*—require 20–40 elements for survival.
- *Carbon*—is an essential part of all organic compounds; photosynthesis is a source of carbon.
- *Photosynthesis*—is the chemical process by which solar energy is stored as chemical energy.

- *Nitrogen*—is required for the construction of proteins and nucleic acids; the major source is the atmosphere.
- *Phosphorus cycle*—is a very inefficient cycle; the greatest source is the lithosphere. Humans have greatly speeded this cycle through mining.
- *Sulfur cycle*—is a cycle in which elementary sulfur of the lithosphere which is not available to plants and animals unless converted to sulfates, is converted.

3.7 CHAPTER REVIEW QUESTIONS

3.1 Define biogeochemical cycle.

3.2 _____ are those parts of the cycle where the chemical is held in large quantities for long periods of time.

3.3 The length of time a chemical is held in an exchange pool or a reservoir is termed its _____ .

3.4 The average length of time a bioelement remains in a compartment is called the _____.

3.5 Name the three transport mechanisms.

3.6 Biochemical cycles can be _____ and/or _____.

3.7 Name the two types of biogeochemical cycles.

3.8 The three main subdivisions of the global environment are:

3.9 The carbon cycle is based on _____.

3.10 "The most important chemical process on earth":

3.11 In the atmosphere, _____ acts as a _____ heat screen.

3.12 The atmosphere contains 78% by volume of _____.

3.13 Aerial nitrogen is converted into _____ mainly by microorganisms, bacteria, and blue-green algae.

3.14 Excretion by animals and anaerobic decomposition of dead organic matter by bacteria produce _____.

3.15 The process whereby ammonia converted by nitrification bacteria into nitrites and then into nitrates is known as _____.

3.16 The process whereby nitrate is converted into nitrogen gas is known as _____.

3.17 The ultimate source of phosphorus is _____.

3.18 Phosphate, upon entering a stream, acts as _____, which promotes the growth of _____.

3.19 Explain the "one-factor control hypothesis."

3.20 The _____ cycle is both sedimentary and gaseous.

CHAPTER 4

Energy Flow in the Ecosystem

The original source of all energy going into food is the sun. This is because plants that have chlorophyll are able to combine water and carbon dioxide in the presence of light energy and produce sugar. This sugar can be converted to energy as the plant needs it. Of course, some of the sugar is converted into other complex chemicals that permit growth, reproduction, and other life processes.[48]

4.1 INTRODUCTION

THE main concepts in this chapter concern how energy moves through an ecosystem. This is important in understanding how ecosystems are balanced, how they may be affected by human activities, and how pollutants will move through an ecosystem.

4.2 FLOW OF ENERGY: THE BASICS

Simply defined, energy is the ability or capacity to do work. For an ecosystem to exist, it must have energy. All activities of living organisms involve work, which is the expenditure of energy. This means the degradation of a higher state of energy to a lower state. The flow of energy through an ecosystem is governed by two laws: the first and second laws of thermodynamics.

The first law, sometimes called the conservation law, states that energy may not be created or destroyed. The second law states that no energy transformation is 100% efficient. That is, in every energy transformation, some energy is dissipated as heat. The term entropy is used as a measure of the nonavailability of energy to a system. Entropy increases with an increase in dissipation. Because of entropy, input of energy in any system is higher than the output or work done; thus, the resultant, efficiency, is less than 100%.

---

[48]Tomera, A. N., *Understanding Basic Ecological Concepts.* Portland, ME: J. Weston Walch, Publisher, p. 43, 1989.

Odum explains that "the interaction of energy and materials in the ecosystem is a primary concern of ecologists."[49] In Chapter 3, biogeochemical nutrient cycles were discussed. It is important to remember that it is the flow of energy that drives these cycles. Moreover, it should be noted that energy does not cycle as nutrients do in biogeochemical cycles. For example, when food passes from one organism to another, energy contained in the food is reduced step by step until all the energy in the system is dissipated as heat (see Section 4.3). Price refers to this process as "a *unidirectional flow* of energy through the system, with no possibility for recycling of energy."[50] When water or nutrients are recycled, energy is required. The energy expended in this recycling is not recyclable. And, as Odum points out, this is a "fact not understood by those who think that artificial recycling of man's resources is somehow an instant and free solution to shortages."[51]

The principal source of energy for any ecosystem is sunlight. Green plants, through the process of photosynthesis, transform the sun's energy into carbohydrates that are consumed by animals. This transfer of energy, as stated previously, is unidirectional—from producers to consumers. Often, this transfer of energy to different organisms is called a food chain. Figure 4.1 shows a simple aquatic food chain.

All organisms, alive or dead, are potential sources of food for other organisms. All organisms that share the same general type of food in a food chain are said to be at the same trophic level (nourishment or feeding level). Because green plants use sunlight to produce food for animals, they are called the producers (the first trophic level). The herbivores, which eat plants directly, are called the second trophic level or the primary consumers. The carnivores are flesh-eating consumers; they include several trophic levels from the third on up. At each transfer, a large amount of energy (about 80 to 90%) is lost as heat and waste. Thus, nature normally limits food chains to four or five links. However, in aquatic ecosystems, "food chains are commonly longer than those of land."[52] The aquatic food chain is longer because several predatory fish may be feeding on the plant consumers. Even so, the built-in inefficiency of the energy transfer process prevents development of extremely long food chains.

Only a few simple food chains are found in nature. Most simple food chains

Figure 4.1 Aquatic food chain.

---

[49]Odum, E. P., *Ecology: The Link Between the Natural and the Social Sciences.* New York: Holt, Rinehart and Winston, Inc., p. 61, 1975.
[50]Price, P. W., *Insect Ecology.* New York: John Wiley & Sons, Inc., p. 11, italics in original, 1984.
[51]Odum, E. P., *Ecology: The Link Between the Natural and the Social Sciences.* New York: Holt, Rinehart and Winston, Inc., p. 61, 1975.
[52]Dasmann, R. F., *Environmental Conservation.* New York: John Wiley & Sons, Inc., p. 65, 1984.

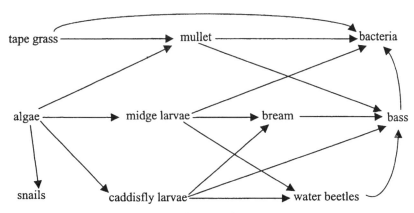

Figure 4.2 Aquatic food web.

are interlocked. This interlocking of food chains forms a food web. A food web can be characterized as a map that shows what eats what.[53] Most ecosystems support a complex food web. A food web involves animals that do not feed on one trophic level. For example, humans feed on plants and animals. An organism in a food web may occupy one or more trophic levels. Trophic level is determined by an organism's role in its particular community, not by its species. Food chains and webs help to explain how energy moves through an ecosystem.

An important trophic level of the food web that has not been discussed thus far is comprised of the decomposers. The decomposers feed on dead plants or animals and play an important role in recycling nutrients in the ecosystem. As Miller points out, "there is no waste in ecosystems. All organisms, dead or alive, are potential sources of food for other organisms."[54] An example of an aquatic food web is shown in the line diagram in Figure 4.2.

4.3 FOOD CHAIN EFFICIENCY

Energy from the sun is captured (via photosynthesis) by green plants and used to make food. Most of this energy is used to carry on the plant's life activities. The rest of the energy is passed on as food to the next level of the food chain.

✓ *Note:* A food chain is the path of food from a given final consumer back to a producer.

---

[53]Miller, G. T., *Environmental Science: An Introduction*. Belmont, CA: Wadsworth, p. 60, 1988.
[54]Miller, G. T., *Environmental Science: An Introduction*. Belmont, CA: Wadsworth, p. 62, 1988.

It is important to note that nature limits the amount of energy that is accessible to organisms within each food chain. Not all food energy is transferred from one trophic level to the next. For ease of calculation, "ecologists often assume an ecological efficiency of 10% (10% rule) to estimate the amount of energy transferred through a food chain."[55] For example, if we apply the 10% rule to the diatoms-copepods-minnows-medium fish-large fish food chain shown in Figure 4.3, we can predict that 1000 grams of diatoms produce 100 grams of copepods, which will produce 10 grams of minnows, which will produce 1 gram of medium fish, which, in turn, will produce 0.1 gram of large fish. Thus, only about 10% of the chemical energy available at each trophic level is transferred and stored in usable form at the next level. The other 90% is lost to the environment as low-quality heat in accordance with the second law of thermodynamics.

✓ *Note:* When an organism loses heat, it represents one-way flow of energy out of the ecosystem. Plants only absorb a small part of energy from the sun, half of which is stored as energy and half of which is lost. The energy plants lose is metabolic heat. Energy from a primary source will flow in one direction through two different types of food chains. In a grazing food chain, the energy will flow from plants (producers) to herbivores and then through some carnivores. In detritus-based food chains, energy will flow from plants through detrivores and decomposers. In terms of the weight (or biomass) of animals in many ecosystems, more of their body mass can be traced to detritus than to living producers. Most of the time, the two food webs will intersect. For example, in the Chesapeake Bay, bass fish of the grazing food web will eat a crab of the detrital food web.[56]

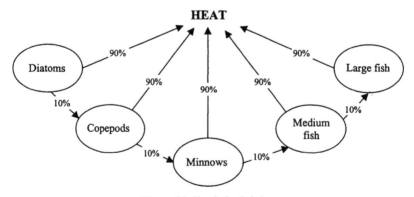

Figure 4.3 Simple food chain.

---

[55]Moran, J. M., Morgan, M. D., and Wiersma, J. H., *Introduction to Environmental Science.* New York: W.H. Freeman and Company, p. 37, 1986.
[56]*Energy Flow Through Ecosystems,* http://clab.cecil.cc.md.us/faculty/biology/Jason/ef.htm, p. 1, 2000.

4.4 ECOLOGICAL PYRAMIDS

As we proceed in the food chain from the producer to the final consumer, it becomes clear that a particular community in nature often consists of several small organisms associated with a smaller number of larger organisms. A grassy field, for example, has a larger number of grass and other small plants, a smaller number of herbivores like rabbits, and an even smaller number of carnivores like foxes. There must be more producers than consumers.

This pound-for-pound relationship, where it takes more producers than consumers, can be demonstrated graphically by building an ecological pyramid. In an ecological pyramid, the number of organisms at various trophic levels in a food chain are represented by separate levels or bars placed one above the other with a base formed by producers and the apex formed by the final consumer. The pyramid shape is formed due to a great amount of energy loss at each trophic level. The same is true if numbers are substituted by the corresponding biomass or energy. Ecologists generally use three types of ecological pyramids: pyramids of number, biomass, and energy. Obviously, there will be differences among them. Some generalizations are as follows:

(1) Energy pyramids must always be larger at the base than at the top (because of the second law of thermodynamics and of dissipation of energy as it moves from one trophic level to another).

(2) Likewise, biomass pyramids (in which biomass is used as an indicator of production) are usually pyramid shaped. This is particularly true of terrestrial systems and aquatic ones dominated by large plants (marshes), in which consumption by heterotroph is low, and organic matter accumulates with time.

Biomass pyramids can sometimes be inverted. This is especially common in aquatic ecosystems, in which the primary producers are microscopic planktonic organisms that multiply very rapidly, have very short life spans, and undergo heavy grazing by herbivores. At any single point in time, the amount of biomass in primary producers is less than that in larger, long-lived animals that consume primary producers.

(3) Numbers pyramids can have various shapes depending on the sizes of the organisms that make up the trophic levels. In forests, the primary producers are large trees, and the herbivore level usually consists of insects, so the base of the pyramid is smaller than the herbivore level above it. In grasslands, the number of primary producers (grasses) is much larger than that of the herbivores above (large grazing animals).[57]

To get a better idea of how an ecological pyramid looks and how it provides

---

[57] *Ecosystem Topics*, http://mason.gmu.edu/~lrockwoo/ECOSYSTEM%20TOPICS.htm, p. 4, 2000.

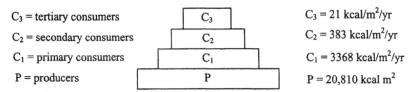

Figure 4.4 Energy flow pyramid. (Source: Adapted from E. P. Odum, *Fundamentals of Ecology*, Philadelphia: Saunders College Publishing, p. 80, 1971.)

its information, we need to look at an example. The example to be used here is the energy pyramid. According to Odum, the energy pyramid is a fitting example because among the "three types of ecological pyramids, the energy pyramid gives by far the best overall picture of the functional nature of communities."[58]

In an experiment conducted in Silver Springs, Florida, Odum measured the energy for each trophic level in terms of kilocalories. A kilocalorie is the amount of energy needed to raise 1 cubic centimeter of water 1°C. When an energy pyramid is constructed to show Odum's findings, it takes on the typical upright form (as it must because of the second law of thermodynamics) as shown in Figure 4.4.

Simply put, as reflected in Figure 4.4 and according to the second law of thermodynamics, no energy transformation process is 100% efficient. This fact is demonstrated, for example, when a horse eats hay. The horse cannot obtain, for his own body, 100% of the energy available in the hay. For this reason, the energy productivity of the producers must be greater than the energy production of the primary consumers. When human beings are substituted for the horse, it is interesting to note that according to the second law of thermodynamics, only a small population could be supported. But, this is not the case. Humans also feed on plant matter, which allows a larger population. Therefore, if meat supplies become scarce, more plant matter must be eaten. This is the situation we see today in countries where meat is scarce. Consider this, if all humans ate soybean, there would be at least enough food for 10 times as many people as compared to a world where everyone eats beef (or pork, fish, chicken, etc.). To demonstrate, every time someone eats meat, food is being taken out of the mouths of nine other people, who could be fed with the plant material that was fed to the animal being eaten.[59] It's not quite that simple, of course, but this is the general idea.

4.5 PRODUCTIVITY

As mentioned previously, the flow of energy through an ecosystem starts

---

[58]Odum, E. P., *Basic Ecology*. Philadelphia: Saunders College Publishing, p. 154, 1983.
[59]*Environmental Biology—Ecosystem*, http://www.marietta.edu/biol.102/ecosystem.html, p. 6, January 10, 1999.

with the fixation of sunlight by plants through photosynthesis. In evaluating an ecosystem, the measurement of photosynthesis is important. Ecosystems may be classified into highly productive or less productive. Therefore, the study of ecosystems must involve some measure of the productivity of that ecosystem.

Smith defines production (or more specifically, primary production, because it is the basic form of energy storage in an ecosystem) as being "the energy accumulated by plants."[60] Primary production is the rate at which the ecosystem's primary producers capture and store a given amount of energy in a specified time interval. In even simpler terms, primary productivity is a measure of the rate at which photosynthesis occurs. Odum lists four successive steps in the production process as follows:

(1) *Gross primary productivity*—is the total rate of photosynthesis in an ecosystem during a specified interval.
(2) *Net primary productivity*—is the rate of energy storage in plant tissues in excess of the rate of aerobic respiration by primary producers.
(3) *Net community productivity*—is the rate of storage of unused organic matter.
(4) *Secondary productivity*—is the rate of energy storage at consumer levels.[61]

When attempting to comprehend the significance of the term productivity as it relates to ecosystems, consider the example of the productivity of an agricultural ecosystem such as a wheat field. Often its productivity is expressed as the number of bushels produced per acre. This is an example of the harvest method for measuring productivity. For a natural ecosystem, several one-square-meter plots are marked off, and the entire area is harvested and weighed to give an estimate of productivity as grams of biomass per square meter per given time interval. From this method, a measure of net primary production (net yield) can be measured.

Productivity, in natural and cultured ecosystems, may vary considerably, not only between type of ecosystems, but also within the same ecosystem. Several factors influence year-to-year productivity within an ecosystem. Such factors as temperature, availability of nutrients, fire, animal grazing and human cultivation activities are directly or indirectly related to the productivity of a particular ecosystem.

The ecosystem that is of greatest importance in this particular study is the aquatic ecosystem. Productivity can be measured in several different ways in the aquatic ecosystem. For example, the production of oxygen may be used to determine productivity. Oxygen content may be measured in the water every few hours for a period of 24 hours. During daylight, when photosynthesis is occurring, the oxygen concentration should rise. At night the oxygen level should

---

[60] Smith, R. L., *Ecology and Field Biology*. New York: Harper & Row, p. 38, 1974.
[61] Odum, E. P., *Fundamentals of Ecology*. Philadelphia: Saunders College Publishing, p. 43, 1971.

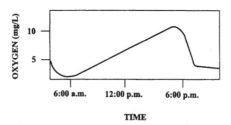

Figure 4.5 The diurnal oxygen curve for an aquatic ecosystem.

drop. The oxygen level can be measured using a simple x-y graph. The oxygen level can be plotted on the y-axis with time plotted on the x-axis, as shown in Figure 4.5.

Another method of measuring oxygen production in aquatic ecosystems is to use light and dark bottles. Biochemical oxygen demand (BOD) bottles (300 ml) are filled with water to a particular level. One of the bottles is tested for the initial dissolved oxygen (DO), then the other two bottles (one clear, one dark) are suspended in the water at the depth from which they were taken. After a twelve-hour period, the bottles are collected, and the DO values for each bottle are recorded. Once the oxygen production is known, the productivity in terms of grams/m/day can be calculated.

In the aquatic ecosystem, pollution can have a profound impact on the system's productivity. For example, certain kinds of pollution may increase the turbidity of the water. This increase in turbidity causes a decrease in energy delivered by photosynthesis to the ecosystem. Accordingly, this turbidity and its aggregate effects decrease net community productivity on a large scale.[62]

4.6 PRODUCTIVITY: THE BOTTOM LINE

Ecological productivity is limping behind human consumption. Since 1984, the global fish harvest has been dropping and so has the per capita yield of grain crops.[63] Moreover, stratospheric ozone is being depleted, the release of greenhouse gases has changed the atmospheric chemistry and might lead to climate change; erosion and desertification is reducing nature's biological productivity; irrigation water tables are falling; contamination of soil and water is jeopardizing the quality of food; other natural resources are being consumed faster than they can regenerate; and biological diversity is being lost. These trends indicate a decline in the quantity and productivity of nature's assets.[64]

---

[62]Laws, E. A., *Aquatic Pollution: An Introductory Text*. New York: John Wiley & Sons, Inc., p. 69, 1993.
[63]Brown, L. R., "Facing food insecurity." In Brown et al., *State of the World*. New York: W.W. Norton, pp. 179–187, 1994.
[64]Wackernagel, M., *Framing the Sustainability Crisis: Getting from Concern to Action*. http://www.sdri.ubc.ca/publications/wachernag.html, p. 4, 1997.

4.7 SUMMARY OF KEY TERMS

- *Energy*—is the ability or capacity to do work. Energy is degraded from a higher to a lower state.
- *First law of thermodynamics*—states that energy is transformed from one form to another but is neither created nor destroyed. Given this principle, we should be able to account for all of the energy in a system in an energy budget, a diagrammatic representation of the energy flows through an ecosystem.
- *Second law of thermodynamics*—asserts that energy is only available due to degradation of energy from a concentrated form to a dispersed form. This indicates that energy becomes more dissipated (randomly arranged) as it is transformed from one form to another or moved from one place to another. It also suggests that any transformation of energy will be less than 100% efficient (i.e., the transfers of energy from one trophic level to another are not perfect; some energy is dissipated during each transfer).
- *Ecological pyramids*—include the pyramids of numbers, productivity, and energy. These pyramids are based on the fact that due to energy loss, fewer animals can be supported at each additional trophic level.

4.8 CHAPTER REVIEW QUESTIONS

4.1 Define the conservation law.

4.2 The primary concern of ecologists is the interaction of _____ and _____ in the ecosystem.

4.3 Define unidirectional flow of energy in an ecosystem.

4.4 The transfer of energy to different organisms is called a _____.

4.5 Interlocked food chains are called a _____.

4.6 _____ feed on dead plants or animals and play an important role in recycling nutrients in the ecosystem.

4.7 Explain the 10% percent rule.

4.8 Explain an ecological pyramid.

4.9 What are the three types of ecological pyramids?

4.10 Define production.

4.11 The rate of storage of unused organic matter is known as _____.

4.12 Net yield is the same as _____.

CHAPTER 5

Population Ecology

The Earth is one but the world is not. We all depend on one biosphere for sustaining our lives. Yet each community, each country, strives for survival and prosperity with little regard for its impact on others. Some consume the Earth's resources at a rate that would leave little for future generations. Others, many more in number, consume far too little and live with the prospects of hunger, squalor, disease, and early death.[65]

5.1 THE 411 ON POPULATION ECOLOGY

LET'S begin with the basics.

5.1.1 POPULATION

Webster's Third New International Dictionary defines population as follows:

- "The total number or amount of things especially within a given area."
- "The organisms inhabiting a particular area or biotype."
- "A group of interbreeding biotypes that represents the level of organization at which speciation begins."

5.1.2 POPULATION SYSTEM

Population system, or life system, is a population with its effective environment.[66]

---

[65]WCED, World Commission on Environment and Development. *Our Common Future.* New York: Oxford University Press, p. 27, 1987.
[66]Clark, L. R., Geier, P. W., Hughes, R. D., and Morris, R. F., *The Ecology of Insect Populations.* New York: Methuen, p. 73, 1967; Berryman, A. A., *Population Systems: A General Introduction.* New York: Plenum, p. 89; Sharov, A. A., "Life-system approach: A systems paradigm in population ecology." *Oikos,* 63: 485–494, 1992.

Major components of a population system are as follows:

(1) *The Population:* organisms in the population can be subdivided into groups according to their age, stage, sex, and other characteristics.
(2) *Resources:* food, shelters, nesting places, space, etc.
(3) *Enemies:* predators, parasites, pathogens, etc.
(4) *Environment:* air, water, soil, temperature, composition, variability of these characteristics in time and space.[67]

5.1.3 POPULATION ECOLOGY[68]

Population ecology is the branch of ecology that studies the structure and dynamics of populations. Population ecology relative to other ecological disciplines is shown in Figure 5.1.

The term "population" is interpreted differently in various sciences. For example, in human demography a population is a set of humans in a given area. In genetics, a population is a group of interbreeding individuals of the same species, which is isolated from other groups. In population ecology, a population is a group of individuals of the same species inhabiting the same area.

✓ *Note: The main axiom of population ecology* is that organisms in a population are ecologically equivalent. Ecological *equivalency* means the following:

(1) Organisms undergo the same life cycle.
(2) Organisms in a particular stage of the life cycle are involved in the same set of ecological processes.
(3) The rates of these processes (or the probabilities of ecological events) are basically the same if organisms are put into the same environment (however, some individual variation may be allowed).[69]

5.2 POPULATION ECOLOGY: HOW IS IT APPLIED TO STREAM ECOLOGY?

If stream ecology students wanted to study the organisms in a slow-moving stream or stream pond, they would have two options. They could study each

---

[67]Sharov, A., *Population Ecology.* http://www.gypsymoth.ento.vt.edu/sharov/popechome/welcome.html, p. 1, 1997.
[68]Sharov, A., *What is Population Ecology?* Blacksburg, VA: Department of Entomology, Virginia Tech University, pp. 1–2, 1996.
[69]WCED, World Commission on Environment and Development. *Our Common Future.* New York: Oxford University Press, p. 2, 1987.

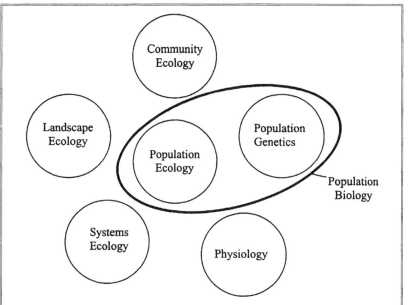

Population ecology—the branch of ecology that studies the structure and dynamics of populations.

Physiology—the study of individual characteristics and individual processes. Used as a basis for prediction of processes at the population level.

Community ecology—the study of the structure and dynamics of animal and plant communities. Population ecology provides modeling tools that can be used for predicting community structure and dynamics.

Population genetics—the study of gene frequencies and micro evolution in populations. Selective advantages depend on the success of organisms in their survival, reproduction, and competition. These processes are studied in population ecology. Population ecology and population genetics are often considered together and called "population biology." Evolutionary ecology is one of the major topics in population biology.

Systems ecology—a relatively new ecological discipline that studies interaction of human population with environment. Major concepts include optimization of ecosystem exploitation and sustainable ecosystem management.

Landscape ecology—another relatively new area in ecology. It studies regional large-scale ecosystems with the aid of computer-based geographic information systems. Population dynamics can be studied at the landscape level, and this is the link between landscape and population ecology.

Figure 5.1 Population ecology relative to other ecological disciplines. (Source: Adapted from Alexi Sharov, *What is Population Ecology?* Blacksburg, VA: Department of Entomology, Virginia Tech University, p. 1, 1966.)

fish, aquatic plant, crustacean, and insect one by one. In that case, they would be studying individuals. It would be easier to do this if the subject were trout, but it would be difficult to separate and study each aquatic plant.

The second option would be to study all of the trout, all of the insects of each specific kind, and all of a certain aquatic plant type in the stream or pond at the time of the study. When stream ecologists study a group of the same kind of individuals in a given location at a given time, they are investigating a population. "Alternately, a population may be defined as a cluster of individuals with a high probability of mating with each other compared to their probability of mating with a member of some other population."[70] When attempting to determine the population of a particular species, it is important to remember that time is a factor. Whether it be at various times during the day, during the different seasons, or from year to year, time is important because populations change.

When measuring populations, the level of species or density must be determined. Density (D) can be calculated by counting the number of individuals in the population (N) and dividing this number by the total units of space (S) the counted population occupies. Thus, the formula for calculating density becomes:

$$D = N/S \qquad (5.1)$$

When studying aquatic populations, the occupied space (S) is determined by using length, width, and depth measurements. The volumetric space is then measured in cubic units.

Population density may change dramatically. For example, if a dam is closed off in a river midway through spawning season, with no provision allowed for fish movement upstream (a fish ladder), it would drastically decrease the density of spawning salmon upstream. Along with the swift and sometimes unpredictable consequences of change, it can be difficult to draw exact boundaries between various populations. Pianka makes this point in his comparison of European starlings that were introduced into Australia with starlings that were introduced into North America. He points out that these starlings are no longer exchanging genes with each other; thus, they are separate and distinct populations.[71]

The population density or level of a species depends on natality, mortality, immigration, and emigration. Changes in population density are the result of births and deaths. The birth rate of a population is called *natality*, and the death rate is called *mortality*. In aquatic populations, two factors besides natality and mortality can affect density. For example, in a run of returning salmon to their spawning grounds, the density could vary as more salmon migrated in or as oth-

---

[70]Pianka, E. R., *Evolutionary Ecology*. New York: HarperCollins, p. 125, 1988.
[71]Pianka, E. R., *Evolutionary Ecology*. New York: HarperCollins, p. 69, 1988.

ers left the run for their own spawning grounds. The arrival of new salmon to a population from other places is termed *immigration* (ingress). The departure of salmon from a population is called *emigration* (egress). Thus, natality and immigration increase population density, whereas mortality and emigration decrease it. The net increase in population is the difference between these two sets of factors.

5.3 DISTRIBUTION

Each organism occupies only those areas that can provide for its requirements, resulting in an irregular distribution. How a particular population is distributed within a given area has considerable influence on density. As shown in Figure 5.2, organisms in nature may be distributed in three ways.

In a random distribution, there is an equal probability of an organism occupying any point in space, and "each individual is independent of the others."[72]

In a regular or uniform distribution, in turn, organisms are spaced more evenly; they are not distributed by chance. Animals compete with each other and effectively defend a specific territory, excluding other individuals of the same species. In regular or uniform distribution, the competition between individuals can be quite severe and antagonistic to the point where the spacing generated is quite even.[73]

The most common distribution is the contagious or clumped distribution where organisms are found in groups; this may reflect the heterogeneity of the habitat. Smith points out that contagious or clumped distributions "produce aggregations, the result of response by plants and animals to habitat differences."[74]

Organisms that exhibit a contagious or clumped distribution may develop social hierarchies in order to live together more effectively. Animals within the same species have evolved many symbolic aggressive displays that carry

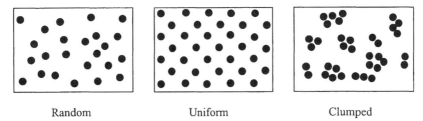

Random   Uniform   Clumped

Figure 5.2 Basic patterns of distribution. (Source: Adapted from E. P. Odum, *Fundamentals of Ecology*. Philadelphia: Saunders College Publishing, p. 205, 1971.)

[72]Smith, R. L., *Ecology and Field Biology*. New York: Harper & Row, p. 292, 1974.
[73]Odum, E. P., *Basic Ecology*. Philadelphia: Saunders College Publishing, p. 97, 1983.
[74]Smith, R. L., *Ecology and Field Biology*. New York: Harper & Row, p. 293, 1974.

meanings that are not only mutually understood but also prevent injury or death within the same species. For example, in some mountainous regions, dominant male bighorn sheep force the juvenile and subordinate males out of the territory during breeding season.[75] In this way, the dominant male gains control over the females and need not compete with other males.

5.4 POPULATION GROWTH

The size of animal populations is constantly changing due to natality, mortality, emigration, and immigration. As mentioned, the population size will increase if the natality and immigration rates are high. On the other hand, it will decrease if the mortality and emigration rates are high. Each population has an upper limit on size, often called the *carrying capacity*. Carrying capacity can be defined as the "optimum number of species' individuals that can survive in a specific area over time."[76] Stated differently, the carrying capacity is the maximum number of species that can be supported in a bioregion. A pond may be able to support only a dozen frogs depending on the food resources for the frogs in the pond. If there were thirty frogs in the same pond, at least half of them would probably die, because the pond environment wouldn't have enough food for them to live. Carrying capacity is based on the quantity of food supplies, the physical space available, the degree of predation, and several other environmental factors.

There are two types of carrying capacity: ultimate and environmental. *Ultimate carrying capacity* is the theoretical maximum density; that is, the maximum number of individuals of a species in a place that can support itself without rendering the place uninhabitable. The *environmental carrying capacity* is the actual maximum population density that a species maintains in an area. Ultimate carrying capacity is always higher than environmental.

Certain species may exhibit several types of population growth. Smith points out that "the rate at which the population grows can be expressed as a graph of the numbers in the population against time."[77] Figure 5.3 shows one type of growth curve.

The J-shaped curve shown in Figure 5.3 shows a rapid increase in size or exponential growth. Eventually, the population reaches an upper limit where exponential growth stops. The exponential growth rate is usually exhibited by organisms that are introduced into a new habitat, by organisms with a short life span such as insects, and by annual plants. A classic example of exponential growth by an introduced species is the reindeer transported to Saint Paul Island

---

[75]Hickman, C. P., Roberts, L. S., and Hickman, F. M., *Biology of Animals.* St. Louis: Times Mirror/Mosby College Publishing, p. 112, 1990.
[76]Enger, E., Kormelink, J. R., Smith, B. F. and Smith, R. J., *Enviromental Science: The Study of Interrelationships.* Dubuque, IA: William C. Brown Publishers, p. 160, 1989.
[77]Smith, R. L., *Ecology and Field Biology.* New York: Harper & Row, p. 317, 1974.

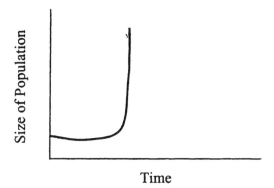

Figure 5.3 J-shaped growth curve.

in the Pribolofs off Alaska in 1911. A total of 25 reindeer were released on the island, and by 1938, there were over 2000 animals on the small island. As time went by, however, the reindeer overgrazed their food supply and the population decreased rapidly. By 1950, only eight reindeer could be found.[78]

Another type of growth curve is shown in Figure 5.4. This logistic or S-shaped (sigmoidal) curve is used for populations of larger organisms having a longer life span. This type of curve has been successfully used by ecologists and biologists to model populations of several different types of organisms, including water fleas, pond snails, and sheep, to name only a few.[79] The curve suggests an early exponential growth phase, while conditions for growth are optimal. As the number of individuals increases, the limits of the environment,

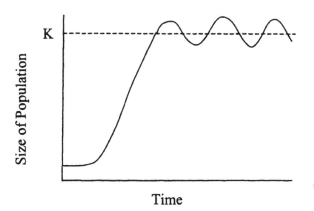

Figure 5.4 S-shaped (sigmoidal) growth curve.

[78]Pianka, E. R., *Evolutionary Ecology*. New York: HarperCollins, p. 163, 1988.
[79]Masters, G. M., *Introduction to Environmental Engineering & Science*. Englewood Cliffs, NJ: Prentice Hall, p. 33, 1991.

or environmental resistance, begin to decrease the number of individuals, and the population size levels off near the carrying capacity, shown as K in Figure 5.4. Usually there is some oscillation around K before the population reaches a stable size, as indicated on the curve.

Mathematically, the S-shaped curve in Figure 5.4 is derived from the following differential equation:

$$Dn/dt = Rn(1 - N/K) \tag{5.2}$$

where N is population size, R is a growth rate, and K is the carrying capacity of the environment. The factor $(1 - N/K)$ is the environmental resistance. As population grows, the resistance to further population growth continuously increases.

It is interesting to note that the S-shaped curve can also be used to find the maximum rate that organisms can be removed without reducing the population size. This concept in population biology is called the *maximum sustainable yield value* of an ecosystem. For example, imagine fishing steelhead fish from a stream. If the stream is at its carrying capacity, theoretically, there will be no population growth, so that any steelheads removed will reduce the population. Thus, the maximum sustainable yield will correspond to a population size less than the carrying capacity. If population growth is logistic or S-shaped, the maximum sustainable yield will be obtained when the population is half the carrying capacity. This can be seen in the following:

The slope of the logistic curve is given by

$$Dn/dt = Rn(1 - N/K)$$

Setting the derivative to zero gives

$$d/dt(Dn/dt) = rdn/dt - rk(2NDn/dt) = 0$$

yielding

$$1 - 2N/K = 0$$

$$N = K/2$$

The logistic growth curve is said to be density conditioned. As the density of individuals increases, the growth rate of the population declines.

As stated previously, after reaching environmental carrying capacity, population normally oscillates around the fixed axis due to various factors that affect the size of the population. These factors work against maintaining population at the K level due to direct dependence on resource availability. *Population controlling factors* affect the size of populations. They are usually grouped into

TABLE 5.1. Factors Affecting Population Size.

Density Independent	Density Dependent
drought	food
fire	pathogens
heavy rain	predators
pesticides	space
human destruction of habitat	psychological disorders
	physiological disorders

two classes, density dependent and density independent. Table 5.1 shows factors that affect population size.

Density-dependent factors are those that increase in importance as the size of the population increases. For example, as the size of a population grows, food and space may become limited. The population has reached the carrying capacity. When food and space become limited, growth is suppressed by competition. Odum describes density-dependent factors as acting "like governors on an engine and for this reason are considered one of the chief agents in preventing overpopulation."[80]

Density-independent factors are those that have the same affect on population regardless of size. Typical examples of density-independent factors are devastating forest fires, streambeds drying up, or the destruction of the organism's entire food supply by disease.

Thus, population growth is influenced by multiple factors. Some of these factors are generated within the population, others from without. Even so, usually no single factor can account fully for the curbing of growth in a given population. It should be noted, however, that humans are, by far, the most important factor; their activities can increase or exterminate whole populations.

5.5 POPULATION RESPONSE TO STRESS

As mentioned earlier, population growth is influenced by multiple factors. When a population reaches its apex of growth (its carrying capacity), certain forces work to maintain population at a certain level. Populations are exposed to small or moderate environmental stresses that work to affect the stability or persistence of the population. Ecologists have concluded that a major factor that affects population stability or persistence is species diversity.

Species diversity is a measure of the number of species and their relative abundance. There are several ways to measure species diversity. One way is to use the straight ratio, $D = S/N$. In this ratio, D = species diversity, N = number of

---

[80]Odum, E. P., *Basic Ecology*. Philadelphia: Saunders College Publishing, p. 339, 1983.

individuals, and S = number of species. As an example, a community of 1000 individuals is counted; the individuals in this community belong to fifty different species. The species diversity would be 50/1000 or 0.050. This calculation does not take into account the distribution of individuals of each species. For this reason, the more common calculation of species diversity is called the Shannon-Weiner Index. The Shannon-Weiner Index measures diversity by

$$H = -\sum_{i=1}^{s} (p_i)(\log p_i) \qquad (5.3)$$

where

H = the diversity index
s = the number of species
i = the species number
p_i = proportion of individuals of the total sample belonging to the ith species

The Shannon-Weiner Index is not universally accepted by ecologists as being the best way to measure species diversity, but it is an example of a method that is available.

Species diversity is related to several important ecological principles. For example, under normal conditions, high species diversity, with a large variety of different species, tends to spread risk. Ecosystems that are in a fairly constant or stable environment, such as a tropical rain forest, usually have higher species diversity. However, as Odum points out, "diversity tends to be reduced in stressed biotic communities."[81]

If the stress on an ecosystem is small, the ecosystem can usually adapt quite easily. Moreover, even when severe stress occurs, ecosystems have a way of adapting. Severe environmental change to an ecosystem can result from natural occurrences such as fires, earthquakes, and floods and from people-induced changes such as land clearing, surface mining, and pollution.

One of the most important applications of species diversity is in the evaluation of pollution. As stated previously, it has been determined that stress of any kind will significantly reduce the species diversity of an ecosystem. In the case of domestic sewage, for example, stress is caused by a lack of dissolved oxygen (DO) for aquatic organisms. This effect is illustrated in Figure 5.5. As illustrated in the graph, the species diversity of a stream exhibits a sharp decline after the addition of domestic sewage.

Ecological succession is the observed process of change (a normal occurrence in nature) in the species structure of an ecological community over time; that is, a gradual and orderly replacement of plant and animal species takes

---

[81]Odum, E. P., *Basic Ecology.* Philadelphia: Saunders College Publishing, p. 409, 1983.

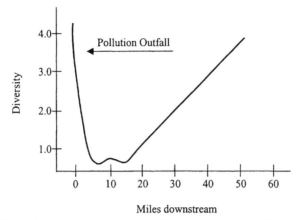

Figure 5.5 Effects of pollution on species diversity. (Source: Adapted from E. P. Odum, *Fundamentals of Ecology*. Philadelphia: Saunders College Publishing, p. 150, 1971.)

place in a particular area over time. For example, if a forest is devastated by a fire, it will grow back, eventually, because of ecological succession.

Additional specific examples of observable succession include the following:

(1) Consider a red pine planting area where the growth of hardwood trees (including ash, poplar, and oak) occurs. The consequence of this hardwood tree growth is increased shading and subsequent mortality of the sun-loving red pines by the shade-tolerant hardwood seedlings. The shaded forest floor conditions generated by the pines prohibit the growth of sun-loving pine seedlings and allow the growth of the hardwoods. The consequence of the growth of the hardwoods is the decline and senescence of the pine forest.

(2) Consider raspberry thickets growing in the sunlit forest sections beneath the gaps in the canopy generated by wind-thrown trees. Raspberry plants require sunlight to grow and thrive. Beneath the dense shade canopy of red pines and oak, there is not sufficient sunlight for the raspberry's survival. However, where a tree has fallen, the raspberry plants proliferate into dense thickets. Within these raspberry thickets, by the way, are dense growths of hardwood seedlings. The raspberry plants generate a protected "nursery" for these seedlings and prevent a major browser of tree seedlings (the white tail deer) from eating and destroying the trees. By providing these trees a shaded haven in which to grow, the raspberry plants are setting up the future tree canopy which will extensively shade the future forest floor and, consequently, prevent the future growth of more raspberry plants!

Succession usually occurs in an orderly, predictable manner that involves the entire system. Ecologists are now able to predict several years in advance what will occur in a given ecosystem. For example, scientists know that if a burned-out forest region receives light, water, nutrients, and an influx or immigration of animals and seeds, it will eventually develop into another forest through a sequence of steps or stages.

Two types of ecological succession are recognized by ecologists: primary and secondary. The particular type that takes place depends on the condition at a particular site at the beginning of the process.

Primary succession, sometimes called *bare-rock succession*, occurs on surfaces such as hardened volcanic lava, bare rock, and sand dunes, where no soil exists, and where nothing has ever grown before (see Figure 5.6). Soil must form on the bare rock before succession can begin. Usually this soil formation process results from weathering. Atmospheric exposure—weathering, wind, rain, and frost—forms tiny cracks and holes in rock surfaces. Water collects in the rock fissures and slowly dissolves the minerals out of the rock's surface. A pioneer soil layer is formed from the dissolved minerals that will support such plants as lichens. Lichens gradually cover the rock surface and secrete carbonic acid, which dissolves additional minerals from the rock. Eventually, the lichens are replaced by mosses. Organisms called decomposers move in and feed on dead lichen and moss. A few small animals such as mites and spiders arrive next. The result is what is known as a *pioneer community*. The pioneer community is defined as the first successful integration of plants, animals, and decomposers into a bare-rock community.[82]

After several years, the pioneer community builds up enough organic matter in its soil to be able to support rooted plants like herbs and shrubs. Eventually, the pioneer community is crowded out and is replaced by a different environment. This, in turn, works to thicken the upper soil layers. The progression continues through several other stages until a mature or climax ecosystem is developed, several decades later. It is interesting to note that in bare-rock succession, each stage in the complex succession pattern dooms the stage that existed before it. According to Tomera, "mosses provide a habitat most inhospitable to lichens, the herbs will eventually destroy the moss community, and so on until the climax stage is reached."[83]

5.5.1 CASE STUDY—FROM LAVA FLOW TO FOREST: PRIMARY SUCCESSION[84]

Probably the best example of primary succession occurred (and is still occurring) on the Hawaiian Islands.

---

[82]Miller, G. T., *Environmental Science: An Introduction.* Belmont, CA: Wadsworth, p. 133, 1988.
[83]Tomera, A. N., *Understanding Basic Ecological Concepts.* Portland, ME: J. Weston Walch, Publisher, p. 66, 1989.
[84]From USGS, *Hawaiian Volcano Observatory,* http://hwo.wr.usgs.gov/volcanowatch/1999/99_01_21.html, p. 1.3, February 1999.

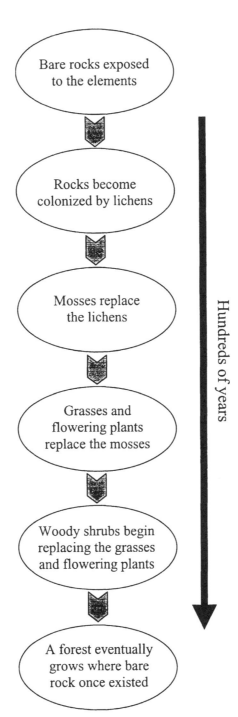

Figure 5.6 Bare-rock succession. (Source: Adapted from A. N. Tomera, *Understanding Basic Ecological Concepts.* Portland, ME: J. Weston Walch, Publisher, p. 67, 1989.)

One of the most striking aspects of a newly formed lava flow is its barren and sterile nature. The process of colonization of new flows begins almost immediately as certain native organisms specially adapted to the harsh conditions begin to arrive from adjoining areas. A wolf spider and cricket may be the first to take up residence, consuming other invertebrates that venture onto the forbidding new environment. The succession process relies heavily on adjacent ecosystems. A steady rain of organic material, seeds, and spores slowly accumulates in cracks and pockets along with tiny fragments of the new lava surface. Some pockets of this infant soil retain enough moisture to support scattered 'ohi'a seedlings and a few hardy ferns and shrubs. Over time, the progeny of these colonizers, and additional species from nearby forests, form an open cover of vegetation, gradually changing the conditions to those more favorable to other organisms. The accumulation of fallen leaves, bark, and dead roots is converted by soil organisms into a thin but rich organic soil. A forest can develop in wet regions in less than 150 years.

On Hawaiian lava flows, primary succession proceeds rapidly on wet windward slopes but more slowly in dry areas. The influence of moisture can be seen on the Kona side, where the same flow can support a forest along the Belt Highway but be nearly barren near the dry coast. Except for the newer flows and disturbed areas, the windward surfaces of Kilauea are heavily forested, but the leeward slope is barren or sparsely vegetated.

All the undisturbed flows on Kilauea, Mauna Loa, and Hualaiai volcanoes are young enough to be in some degree of primary succession, and the patterns and relative ages of lava flows are reflected in the maturity of vegetation. Only a few of the newest flows on the dry upper slopes of dormant Mauna Kea are young enough to reflect primary succession. Extinct Kohala volcano is too old to find such flows, and vegetation differences reflect rainfall amounts and disturbance.

On the wetter slopes of Hualaiai and Mauna Loa, younger flows stand out against a more uniform, older background, as the surfaces are recovered by lava at rates of only 20–40% a century. Small and more active, Kilauea renews about 90% of its surface in the same time period, and the resulting pattern is a patchwork of flows and vegetated remnants (kipuka). The many younger flows rely on the older kipuka to provide sources of plants and animals.

The native forest ecosystems have adapted to the overpowering nature of volcanic eruptions by being able to quickly recolonize from the many kipuka around new flows. However, the added losses due to forest clearing and introduced organisms provide additional threats to which the native biota is not adapted. If too many of the native forest areas are cleared or taken over by introduced organisms, natural succession may not be able to provide a replacement native ecosystem on the younger flows. The continuing primary succession process may already be partially interrupted in low Puna, where so much of the native forest has been cleared for development and where colonizers from nearby areas are mostly introduced organisms.

Secondary succession is the most common type of succession. Secondary succession occurs in an area where the natural vegetation has been removed or destroyed, but the soil is not destroyed. For example, succession that occurs in abandoned farm fields, known as *old field succession,* illustrates secondary succession. An example of secondary succession can be seen in the Piedmont region of North Carolina. Early settlers of the area cleared away the native oak-hickory forests and cultivated the land. In the ensuing years, the soil became depleted of nutrients, reducing the soil's fertility. As a result, farming ceased in the region a few generations later, and the fields were abandoned. Some 150 to 200 years after abandonment, the climax oak-hickory forest was restored.

5.6 POPULATION RESPONSE TO STRESS AND STREAM ECOLOGY

It is important to understand that the dynamic balance of the stream ecosystem is between population growth and population reduction factors. Factors that cause the population to increase in number are growth factors. Factors that cause the population to decrease in number are called reduction factors.

In a stream ecosystem, growth is enhanced by biotic and abiotic factors. These factors include the following:

(1) Ability to produce offspring

(2) Ability to adapt to new environments

(3) Ability to migrate to new territories

(4) Ability to compete with species for food and space to live

(5) Ability to blend into the environment so as not to be eaten

(6) Ability to find food

(7) Ability to defend against enemies

(8) Favorable light

(9) Favorable temperature

(10) Favorable dissolved oxygen (DO) content

(11) Sufficient water level

The biotic and abiotic factors in a stream ecosystem that reduce growth include the following:

(1) Predators

(2) Disease
(3) Parasites
(4) Pollution
(5) Competition for space and food
(6) Unfavorable stream conditions (i.e., low water levels)
(7) Lack of food

✓ *Note:* When all populations within a stream ecosystem are in balance, the entire stream ecosystem is in balance.

In regards to stability in a stream ecosystem, the higher the species diversity, the greater the inertia and resilience of the ecosystem. At the same time, when the species diversity is high within a stream ecosystem, a population within the stream can be out of control because of an imbalance between growth and reduction factors, but the ecosystem can remain stable.

In regards to instability in a stream ecosystem, recall that imbalance occurs when growth and reduction factors are out of balance. For example, when sewage is accidentally dumped into a stream, the stream ecosystem, via the self-purification process (discussed later), responds and returns to normal. This process is described as follows:

(1) Raw sewage is dumped into the stream.
(2) This sewage decreases the oxygen available as the detritus food chain breaks down the sewage.
(3) Some fish die at the pollution site and downstream.
(4) Sewage is broken down, washes out to sea, and is finally broken down in the ocean.
(5) Oxygen levels return to normal.
(6) Fish populations that were deleted are restored as fish reproduce and the young occupy the area formerly occupied by the dead fish.
(7) Populations all return to "normal."

A shift in balance in a stream's ecosystem (or in any ecosystem) similar to the one just described is a fairly common occurrence. In this particular case, the stream responded (on its own) to the imbalance the sewage caused, and through the self-purification process, returned to normal. Recall that we defined succession as being the method by which an ecosystem either forms itself or heals itself. Thus, we can say that a type of succession has occurred in the polluted stream described above, because, in the end, it healed itself.

In summary, through research and observation, ecologists have found that the succession patterns in different ecosystems usually display common characteristics. First, succession brings about changes in the plant and animal members present. Second, organic matter increases from stage to stage. Finally, as each stage progresses, there is a tendency toward greater stability or persistence. Succession is usually predictable unless humans interfere. This illustrates Garrett Hardin's *First Law of Ecology:* We can never do merely one thing. Any intrusion into nature has numerous effects, many of which are unpredictable.[85]

5.7 SUMMARY OF KEY TERMS

- *Population density*—is the number of a particular species in an area. This is affected by *natality* (birth and reproduction), *immigration* (moving into), *mortality* (death), and *emigration* (moving out of).
- *Ultimate carrying capacity*—is the maximum number of a species an area can support; the *environmental carrying capacity* is the actual maximum capacity a species maintains in an area. Ultimate capacity is always greater than environmental capacity.

5.8 CHAPTER REVIEW QUESTIONS

5.1 Define population ecology.

5.2 What is the main axiom of population ecology?

5.3 When measuring populations, the level of _____ or _____ must be determined.

5.4 The arrival of new species to a population from other places is termed _____.

5.5 _____ studies the structure and dynamics of animal and plant communities.

5.6 _____ produces aggregation, the result of responses by plants and animals to habitat differences.

[85]Miller, G. T., *Environmental Science: An Introduction.* Belmont, CA: Wadsworth, p. 233, 1988.

5.7 _____ is the upper limit of population size.

5.8 The _____ is the actual maximum population density that a species maintains in an area.

5.9 _____ factors affect the size of populations.

5.10 _____ is the observed process of change in the species structure of an ecological community over time.

5.11 Describe bare-rock succession.

5.12 The first successful integration of plants, animals, and decomposers into a bare-rock community is called a _____.

5.13 In a stream ecosystem, growth is enhanced by _____ and _____ factors.

5.14 State Hardin's First Law of Ecology.

CHAPTER 6

Stream Water

Our planet is shrouded in water, and yet 8 million children under the age of five will die this year from lack of safe water. The same irony will see 800 million people at risk from drought. . . . Two-thirds of the world's rural poor have no access to safe drinking water, and while millions are made homeless from floods, hundreds of millions are coping with drought.[86]

6.1 WHERE IS EARTH'S WATER LOCATED?

WATER continually moves around and above the earth as water vapor, liquid water, and ice. Moreover, water is continually changing its form. The earth is almost a "closed system," like a terrarium. That means that the earth, as a whole, neither gains nor loses much matter, including water. The water cycle (see Section 6.3) continually recycles water all around the globe, which means that the water we drink today was once used by our predecessors, in near time and very ancient time.[87]

Water covers three-quarters of the earth's surface. Most of this water is saltwater (about 97% of the earth's water) in the oceans, and less than 3% is freshwater. Of the latter, 77% is frozen in polar ice caps and glaciers, 22% is groundwater, and the remaining small fraction is in lakes, rivers, streams, plants, and animals.

From Figure 6.1, it is apparent that we generally only make use of a tiny portion of the available water supplies. The right-side pie shows that the vast majority of the freshwater available for our uses is stored in the ground.

Look at the data provided in Table 6.1. Surface-water sources (such as streams and rivers) only constitute about 300 cubic miles (about 1/10,000th of one percent) of the world's total water supply of about 326 million cubic miles of water; yet streams and rivers are the source of most of the water used.

---

[86]*What Water Shortage?* New York: United Nations Environment Program (UNEP), pp. 1–5, undated.
[87]*Where is Earth's Water Located?* USGS Water Resources: http://www.ga.usga.gov/edu/earthwherewater.html, pp. 1–2, February 2000.

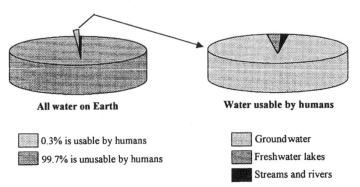

Figure 6.1 Earth's water available for our uses, and the forms in which it exists. (Source: USGS Water Resources, hperlman@usgs.gov, p. 2, 2000.)

Through history, the world's surface waters, i.e., lakes, streams, and rivers, have provided important resources and services (water available for human use), including water for drinking, washing, agriculture, energy production, transportation, recreation, and waste disposal.

Humans cannot live without water. Consider the 5-5-5 Rule—humans can live about five minutes without air, about five days without water, and about five weeks without food. And, water has several uses. All essential "biochemical processes occur in water."[88] Moreover, water has limitations. It is very unevenly distributed around the world. Much of the Middle East, most of Africa, parts of Central America, and the western United States are already short of water.[89] Unfortunately, water has been treated as an unlimited resource. This atti-

TABLE 6.1. Earth's Water Supply.

Water Source	Water Volume, in Cubic Miles	Percent of Total Water
Oceans	317,000,000	97.24%
Ice caps, glaciers	7,000,000	2.14%
Groundwater	2,000,000	0.61%
Freshwater lakes	30,000	0.009%
Inland seas	25,000	0.008%
Soil moisture	16,000	0.005%
Atmosphere	3,100	0.001%
Rivers (streams)	300	0.0001%
Total water volume	326,000,000	100%

Source: Where is Earth's Water Located? USGS Water Resources: http://www.ga.usgs.gov/edu/earthwherewater.html, p. 2, February 2000.

[88] Bradbury, I., *The Biosphere*. New York: Belhaven Press, p. 16, 1991.
[89] *World Resources 1986*. New York: World Resources Institute (WRI) and International Institute of Environment and Development (IIED), pp. 113–116, 1986.

TABLE 6.2. U.S. Water Withdrawals per Day, 1940–1985.

Year	Billion Gallons
1940	140
1950	180
1960	270
1970	370
1980	450
1985	400

Source: U.S. Bureau of Census (phone query, 1996).

tude, if continued, could lead to critical shortages in quantity and critical deficiencies in quality of available water.

Another water limitation concerns usage; that is, the trend toward overusage. For example, from Table 6.2, it can be seen that between 1950 and 1980, the amount of water drawn from lakes, rivers, streams, reservoirs, and underground aquifers in the United States increased by 150%, while the population increased by only half.[90]

Another water limitation is water quality. Data compiled by WRI and IIED indicate that the world's lakes and rivers receive enormous quantities of municipal sewage, industrial discharges, and surface runoff from agricultural areas on a continuous basis. If the discharge of pollutants into our groundwater and surface waters is not stopped, the survival of future generations is at risk.[91]

To this point in this discussion, we have covered the fundamentals of ecology in limited and very basic form. The intent has been to lay a foundation upon which the focus of this text, stream ecology and self-purification, can be built. Beginning with an introduction to water and its importance, the rest of this discussion focuses on the main topic area.

6.2 WATER: EARTH'S BLOOD

Water is the life blood of the universe. Living organisms, themselves about 70% water, depend upon water as a medium and reactant for biochemical reactions, for circulation, and for support. Water covers about 75% of the earth's surface. Life on the earth probably originated in water. More than half of the world's animals and plant species live in water. Most of our food is water: tomatoes (95%), spinach (91%), milk (90%), apples (85%), potatoes (80%), beef (61%), and hot dogs (56%). Water is at once simple and complex. It is so complex that we can't define, beyond its chemical formula, what water really is.[92]

---

[90]Weber, S. (Ed.), *USA by Numbers.* Washington, DC: Zero Population Growth, p. 17, 1988.
[91]WRI and IIED. *World Resources 1988–1989.* New York: World Resources Institute (WRI) and International Institute of Environment and Development (IIED), p. 133, 1988.
[92]Spellman, F. R., *The Science of Water: Concepts and Applications.* Lancaster, PA: Technomic Publishing Company, Inc., p. 13, 1998.

Whether we can define it or not, however, humans know water not so much for what it is, but instead for its use(s). Water is used to maintain health and for industrial and home use, sanitation, agriculture, power production, recreation, and much else.

Water is a stable molecule composed of one atom of oxygen and two atoms of hydrogen. Water is unique in that it is the only material found on earth in the three basic states at standard temperatures. These states are liquid (water), solid (ice), and gas (water vapor). At sea level water vaporizes at 100°C (212°F) and freezes at 0°C (32°F). At 4°C (39.2°F), water is most dense. In a body of water, as water approaches this temperature during the spring and fall, a mixing action takes place whereby the denser water displaces water at lower levels. This mixing action aerates the water and brings nutrients to the surface.

6.3 WATER, OR HYDROLOGIC, CYCLE

The water, or hydrologic, cycle is shown in Figure 6.2. As illustrated, the water cycle depicts the ongoing natural circulation of water through the biosphere. Water is taken from the earth's surface to the atmosphere by evaporation from the surface of lakes, rivers, streams, and oceans. This evaporation process occurs when water is heated by the sun. The sun's heat energizes surface molecules, allowing them to break free of the attractive force binding them together, and then evaporate and rise as invisible vapor in the atmosphere. Water vapor is also emitted from plant leaves by a process called *transpiration*. Every day, an actively growing plant transpires five to ten times as much water as it can hold at once. As water vapor rises, it cools and eventually condenses, usually on tiny particles of dust in the air. When it condenses, it becomes a liquid again or turns directly into a solid (ice, hail, or snow). These water particles then collect and form clouds. The atmospheric water formed in clouds eventually falls to earth as precipitation. Most precipitation falls in coastal areas or in high elevations. Some of the water that falls in high elevations becomes runoff water, the water that runs over the ground (sometimes collecting nutrients from the soil) to lower elevations to form streams, lakes, and fertile valleys.

The water that we see is known as *surface water*. Surface water can be broken down into five categories: oceans, lakes, rivers and streams, estuaries, and wetlands.

6.4 STREAM WATER

As Allen points out, we all have an intuitive appreciation that flowing waters, such as streams, contain a variety of dissolved and suspended constituents (see Figure 6.3).[93] Further, a stream is a carrier of those constituents it is exposed to as it makes its inexorable journey over land to the sea.

---

[93] Allen, J. D., *Stream Ecology: Structure and Function of Running Waters*. London: Chapman & Hall, p. 23, 1996.

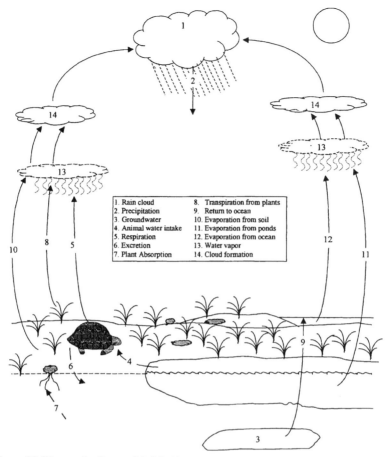

Figure 6.2 Water cycle. (Source: Modified from Carolina Biological Supply Co., 1966, with permission.)

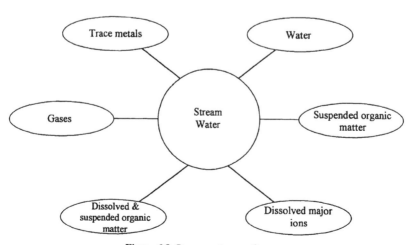

Figure 6.3 Stream water constituents.

69

Many factors influence the composition of stream water, causing variations from place to place. Precipitation is, of course, one major source of chemical inputs to streams. Most streams contain much more suspended and dissolved materials than found in rainwater, snow, or hail, however. Ultimately, all of the constituents of stream water originate from dissolution of the earth's rocks. The dissolving of rocks is commonly the major determinant of stream water chemistry locally as well. These "determinants" of stream water chemistry are actually the stream's "DNA." More specifically, if we were to analyze a sample of stream water, at any given time or place, the dissolved and suspended constituents found in the stream water sample would literally provide us with a sort of "DNA" footprint, or signature, of where the stream has been. It is this "DNA" footprint that allows us to determine and often pinpoint, in many cases, the origin of point- and non-point source pollution.

In our pursuit of presenting a logical discussion of stream ecology and self-purification, we share the view of Dr. David M. Rosenberg: "Chemical measurements are like taking snapshots of the ecosystem, whereas biological measurements are like making a videotape. . . . "[94] Thus, in this discussion of stream ecology and self-purification, it is the videotape (the biological) that we focus on.

6.5 KEY TERM

- *Hydrologic Cycle (Water Cycle)*—involves the circulation of water between the earth's surface and its atmosphere.

6.6 CHAPTER REVIEW QUESTIONS

6.1 Surface-water sources constitute about _____ cubic miles of the world's water supply.

6.2 Less than _____% of the earth's water supply is freshwater.

6.3 The earth is almost a closed system. Explain.

---

[94]Rosenberg, D. M., "A National Aquatic ecosystem health program for Canada: We should go against the flow." *Bull. Entomolo. Soc. Can.*, Winnipeg Freshwater Institute, 30(4):144–152, 1998.

6.4 Explain the 5-5-5 rule.

6.5 Humans know water not so much for what it is, but instead for its _____.

6.6 Explain the hydrologic cycle.

CHAPTER 7

Freshwater Ecology

Ours is a water planet. Water covers three quarters of its surface, makes up two-thirds of our bodies. It is so vital to life we can't live more than four days without it. If all the earth's water—an estimated 325 trillion gallons—were squeezed into a gallon jug and you poured off what was not drinkable (too salty, frozen or polluted) you'd be left with one drop. And even that might not pass U.S. water quality standards.[95]

7.1 NORMAL STREAM LIFE

NORMAL stream life can be compared to that of a "balanced aquarium."[96] That is, nature continuously strives to provide clean, healthy, normal streams. This is accomplished by maintaining the stream's flora and fauna in a balanced state. Nature balances stream life by maintaining both the number and the type of species present in any one part of the stream. Such balance ensures that there is never an overabundance of one species. Nature structures the stream environment so that plant and animal life is dependent upon the existence of others within the stream. Thus, nature has structured an environment that provides for interdependence, which leads to a *balanced aquarium* in a normal stream.

7.2 FRESHWATER ECOLOGY

To this point, the fundamental concepts of ecology, which are generally related to both terrestrial and freshwater habitats, have been discussed. Before narrowing the focus to the topic of freshwater ecology and, more particularly, stream ecology and self-purification, the two different ecosystems, land and freshwater habitats, will be contrasted.

---

[95]Narr, J., *Design for a Livable Planet.* New York: Harper & Row, p. 61, 1990.
[96]ASTM. *Manual on Water.* Philadelphia: American Society for Testing and Materials, p. 86, 1969.

The major difference between land and freshwater habitats is in the medium in which both exist. The land or terrestrial habitat is enveloped in a medium of air, the atmosphere. The freshwater habitat, on the other hand, exists in a water medium. Although the two ecosystems are different, they both use oxygen. Contrast exists in how the oxygen is formulated in each system and how organisms utilize it.

The following data clearly illustrate this contrast. For example, atmospheric air contains at least twenty times more oxygen than does water. Air has approximately 210 ml of oxygen per liter; water contains 3–9 ml per liter, depending on temperature, presence of other solutes, and degree of saturation. Moreover, freshwater organisms must work harder for their oxygen. That is, they must expend far more effort extracting oxygen from water than land animals expend removing oxygen from air.[97]

Other contrasts between land and water ecosystems can be seen in other comparisons. For example, water is approximately 1000 times denser than air and approximately 50 times more viscous. Additionally, natural bodies of water have tremendous thermal capacity, with little temperature fluctuation, as compared to atmospheric air.

Freshwater ecology is the branch of ecology that deals with the biological aspect of *limnology*. Limnology, as defined by Welch, "deals with biological productivity of inland waters and with all the causal influences which determine it."[98] Limnology divides freshwater ecosystems into two groups or classes, lentic and lotic habitats. The *lentic* (Lenis = calm) or standing water habitats are represented by lakes, ponds, and swamps. The *lotic* (Lotus = washed) or running water habitats are represented by rivers, streams, and springs. On occasion, these two different habitats are not well differentiated. This can be seen in the case of an old, wide, and deep river where water velocity is quite low, and the habitat, therefore, becomes similar to that of a pond.

7.3 LENTIC HABITAT

Lakes and ponds range in size of just a few square feet to thousands of square miles. Scattered throughout the earth, many of the first lakes evolved during the Pleistocene Ice Age. Many ponds are seasonal, just lasting a couple of months, such as sessile pools, while lakes last many years. There is not that much diversity in species, because lakes and ponds are often isolated from one another and from other water sources such as streams and oceans.

Lakes and ponds are divided into four different "zones" that are usually determined by depth and distance from the shoreline. The four distinct

---

[97]Hickman, C. P., Roberts, L. S., and Hickman, F. M., *Integrated Principles of Zoology*. St. Louis: Times Mirror/Mosby College Publishing, p. 161, 1988.
[98]Welch, P. S., *Limnology*. New York: McGraw-Hill, p. 10, 1963.

zones—littoral, limnetic, profundal, and benthic—are shown in Figure 7.1. Each zone "provides a variety of ecological niches for different species of plant and animal life."[99]

The *littoral zone* is the topmost zone near the shores of the lake or pond with light penetration to the bottom. It provides an interface zone between the land and the open water of lakes. This zone contains rooted vegetation such as grasses, sedges, rushes, water lilies and waterweeds, and a large variety of organisms. The littoral zone is further divided into concentric zones, with one group replacing another as the depth of water changes. Figure 7.1 also shows these concentric zones—emergent vegetation, floating leaf vegetation, and submerged vegetation zones—proceeding from shallow to deeper water.

The littoral zone is the warmest zone because it is the area that light hits, it contains flora such as rooted and floating aquatic plants, and it contains a very diverse community, which can include several species of algae (like diatoms), grazing snails, clams, insects, crustaceans, fishes, and amphibians. The aquatic plants aid in providing support by establishing "excellent habitats for photosynthetic and heterotrophic (requires organic food from the environment) microflora as well as many zooplankton and larger invertebrates."[100] In the case of insects, such as dragonflies and midges, only the egg and larvae stages are found in this zone. The fauna includes such species as turtles, snakes, and ducks that feed on the vegetation and other animals in the littoral zone.

Figure 7.2 shows a top view of the other zones making up the littoral zone.

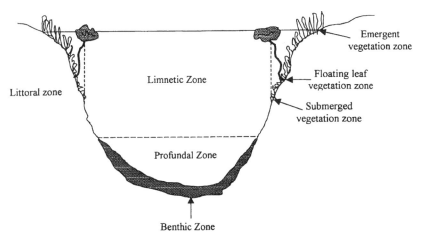

Figure 7.1 Vertical section of a pond showing major zones of life. (Source: Modified from E. Enger, J. R. Kormelink, B. F. Smith, and R. J. Smith, *Environmental Science: The Study of Interrelationships.* Dubuque, IA: William C. Brown Publishers, p. 77, 1989.)

[99]Miller, G. T., *Environmental Science: An Introduction.* Belmont, CA: Wadsworth, p. 77, 1988.
[100]Wetzel, R. G., *Limnology.* Orlando, FL: Harcourt Brace, p. 519, 1983.

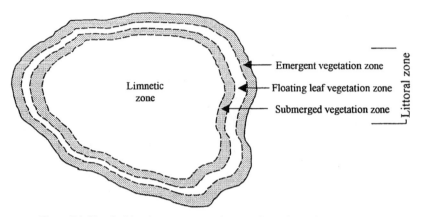

Figure 7.2 View looking down on concentric zones that make up the littoral zone.

From Figure 7.2, it can be seen that the *limnetic zone* is the open-water zone up to the depth of effective light penetration; that is, the open water away from the shore. The community in this zone is dominated by minute suspended organisms, the plankton, such as phytoplankton (plants) and zooplankton (animals), and some consumers such as insects and fish. Plankton are small organisms that can feed and reproduce on their own and serve as food for small chains.

✓ *Note:* Without plankton in the water, there would not be any living organisms in the world, including humans.

In the limnetic zone, the population density of each species is quite low. The rate of photosynthesis is equal to the rate of respiration; thus, the limnetic zone is at compensation level. Small shallow ponds do not have this zone; they have only a littoral zone. When all lighted regions of the littoral and limnetic zones are discussed as one, the term *euphotic* is used, designating these zones as having sufficient light for photosynthesis and the growth of green plants to occur.

The small plankton do not live for a long time. When they die, they fall into the deep-water part of the lake/pond, the *profundal zone*. The profundal zone, because it is the bottom or deep-water region, is not penetrated by light. This zone is primarily inhabited by heterotrophs adapted to its cooler, darker water and lower oxygen levels.

The final zone, the *benthic zone*, is the bottom region of the lake. It supports scavengers and decomposers that live on sludge. The decomposers are mostly large numbers of bacteria, fungi, and worms, which live on dead animal and plant debris and wastes that find their way to the bottom.

7.4 LOTIC HABITAT

Lotic (washed) habitats are characterized by continuously running water or current flow. These running water bodies, rivers and streams, typically have three zones: riffle, run, and pool. The *riffle zone* contains faster-flowing, well-oxygenated water, with coarse sediments. In the riffle zone, the velocity of current is great enough to keep the bottom clear of silt and sludge, thus providing a firm bottom for organisms. This zone contains specialized organisms that are adapted to live in running water. For example, organisms adapted to live in fast streams or rapids (trout) have streamlined bodies, which aid in their respiration and in obtaining food.[101] Stream organisms that live under rocks to avoid the strong current have flat or streamlined bodies. Others have hooks or suckers with which to cling or attach to a firm substrate to avoid the washing-away effect of the strong current.[102]

The *run zone* (or intermediate zone) is the slow-moving, relatively shallow part of the stream with moderately low velocities and little or no surface turbulence.

The *pool zone* of the stream is usually a deeper water region where velocity of water is reduced and silt and other settling solids provide a soft bottom (more homogeneous sediments), which is unfavorable for sensitive bottom-dwellers. Decomposition of some of these solids causes a lower amount of dissolved oxygen (DO). It is interesting to note that some stream organisms spend some of their time in the rapids part of the stream and other times in the pool zone (trout, for example). Trout typically spend about the same amount of time in the rapid zone pursuing food as they do in the pool zone pursuing shelter.

Organisms are sometimes classified based on their modes of life. The following section provides a listing of the various classifications based on mode of life.

7.4.1 CLASSIFICATION OF AQUATIC ORGANISMS BASED ON MODE OF LIFE

(1) *Benthos (Mud Dwellers):* this term originates from the Greek word for bottom and broadly includes aquatic organisms living on the bottom or on submerged vegetation. They live under and on rocks and in the sediments. A shallow sandy bottom has sponges, snails, earthworms, and some insects. A deep, muddy bottom will support clams, crayfish, nymphs of damselflies,

---

[101] Smith, R. L., *Ecology and Field Biology.* New York: Harper & Row, p. 134, 1974.
[102] Allen, J. D., *Stream Ecology: Structure and Function of Running Waters.* London: Chapman & Hall, p. 48, 1996.

dragonflies, and mayflies. A firm, shallow, rocky bottom has nymphs of mayflies, stone flies, and larvae of water beetles.

(2) *Periphytons or Aufwuchs:* the first term usually refers to microfloral growth upon substrata (i.e., benthic-attached algae). The second term, *aufwuchs* (pronounce: OWF-vooks; German: "growth upon"), refers to the fuzzy, sort of furry-looking, slimy green coating that attaches or clings to stems and leaves of rooted plants or other objects projecting above the bottom without penetrating the surface. It consists not only of algae like Chlorophyta, but also diatoms, protozoans, bacteria, and fungi.

(3) *Planktons (Drifters):* they are small, mostly microscopic plants and animals that are suspended in the water column; movement depends on water currents. They mostly float in the direction of the current. There are two types of planktons: *phytoplanktons* are assemblages of small plants (algae) with limited locomotion abilities (they are subject to movement and distribution by water movements) and *zooplankton* are animals that are suspended in water with limited means of locomotion (examples include crustaceans, protozoans, and rotifers).

(4) *Nektons or Pelagic Organisms (capable of living in open waters):* they are distinct from other planktons in that they are capable of swimming independent of turbulence. They are swimmers that can navigate against the current. Examples of nektons include fish, snakes, diving beetles, newts, turtles, birds, and large crayfish.

(5) *Neustons:* they are organisms that float or rest on the surface of the water. Some varieties can spread their legs so that the surface tension of the water is not broken; for example, water striders (see Figure 7.3).

(6) *Madricoles:* organisms that live on rock faces in waterfalls or seepages.

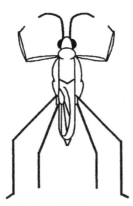

Figure 7.3 Water strider. (Source: *Standard Methods,* 15th Edition. Copyright ©1981 by the American Public Health Association, the American Water Works Association, and the Water Pollution Control Federation; reprinted with permission.)

7.5 LIMITING FACTORS

An aquatic community has several unique characteristics. The aquatic community operates under the same ecologic principles as terrestrial ecosystems, but the physical structure of the community is more isolated and exhibits limiting factors that are very different from the limiting factors of a terrestrial ecosystem.

Certain materials and conditions are necessary for the growth and reproduction of organisms. If, for instance, a farmer plants wheat in a field containing too little nitrogen, it will stop growing when it has used up the available nitrogen, even if the wheat's requirements for oxygen, water, potassium, and other nutrients are met. In this particular case, nitrogen is said to be the limiting factor.

A *limiting factor* is a condition or a substance (the resource in shortest supply) that limits the presence and success of an organism or a group of organisms in an area. There are two well-known laws about limiting factors:

(1) *Liebig's Law of the Minimum:* Odum has modernized Liebig's Law in the following: "Under steady state conditions the essential material available in amounts most closely approaching the critical minimum needed, will tend to be the limiting one."[103] Liebig's Law is normally restricted to chemicals that limit plant growth in the soil, for instance, nitrogen, phosphorus, and potassium. It does not deal with the excess of a factor as limiting.

(2) *Shelford's Law of Tolerance:* although Liebig's Law does not deal with the excess of a factor as limiting, excess is or can be a limiting factor. The presence and success of an organism depends on the completeness of a complex of conditions. Odum describes Shelford's Law of Tolerance as follows: "Absences or failure of an organism can be controlled by the qualitative and quantitative deficiency or excess with respect to any one of the several factors which may approach the limits of tolerance for that organism."[104] For instance, too much and too little heat, light, and moisture can be limiting factors for some plants.

Price points out that "these two laws actually relate to individual organisms, and the survival of an individual in a given set of conditions, independent of others in the same niche."[105] Expressed differently, both of these laws state that the presence and success of an organism or a group of organisms depend upon a complex of conditions, and any condition that approaches or exceeds the limits of tolerance is said to be a limiting condition or factor.

---

[103] Odum, E .P., *Fundamentals of Ecology*. Philadelphia: Saunders College Publishing, p. 106, 1971.
[104] Odum, E .P., *Fundamentals of Ecology*. Philadelphia: Saunders College Publishing, p. 107 1971.
[105] Price, P. W., *Insect Ecology*. New York: John Wiley & Sons, Inc., p. 415, 1984.

Several factors affect biological communities in streams. These include the following:

- water quality
- temperature
- turbidity
- dissolved oxygen
- acidity
- alkalinity
- organic and inorganic chemicals
- heavy metals
- toxic substances
- habitat structure
- substrate type
- water depth and current velocity
- spatial and temporal complexity of physical habitat
- flow regime
- water volume
- temporal distribution of flows
- energy sources
- type, amount, and particle size of organic material entering stream
- seasonal pattern of energy availability
- biotic interactions
- competition
- predation
- disease
- parasitism
- mutualism

The common physical limiting factors in freshwater ecology important to this discussion include the following:

(1) Temperature
(2) Light
(3) Turbidity
(4) Dissolved atmospheric gases, especially oxygen
(5) Biogenic salts in macro- and micronutrient forms
   - macronutrients, such as nitrogen, phosphorus, potassium, calcium, and sulfur
   - micronutrients such as iron, copper, zinc, chlorine, and sodium
(6) Water movement—stream currents, especially rapids

7.5.1 TEMPERATURE

Aquatic organisms are very sensitive to temperature change, as water temperature generally does not change rapidly. It should be noted, however, that surface waters can be subject to great temperature variations. Tchobanoglous and Schroeder point out that across the United States, for example, surface water temperatures can vary from 0.5°C to 27°C.[106] Water has some unique properties, such as very high molar heat of fusion (1.44 kcal) and molar heat of vaporization (9.70 kcal), which allow a very slow change in water temperature.

Aquatic organisms often have narrow temperature tolerance and are known as *stenothermal* (narrow temperature range). The limits for abrupt changes in water temperature are −5°F. Water has its greatest density (1 g/cm^3) at 4°C. Above and below this temperature, it is lighter. Temperature changes, therefore, produce a characteristic pattern of stratification of lakes and ponds in tropical and temperate regions, which helps the aquatic life to survive under severe winter and summer conditions (see Figure 7.4). Figure 7.4 shows the effect temperature has on thermal stratification, which causes *turning over* of lakes and ponds.

During the summer, turning over occurs because the top layer of water becomes warmer than the bottom; and as a result, there are two layers of water, the top one lighter and the bottom one heavier. With the further rise in temperature, the top layer becomes even lighter than the bottom layer, and a middle layer with medium density is created. These layers, from top to bottom, are known as *epilimnion, thermocline*, and *hypolimnion*. They are lightest and warmest, medium weight and warmer, and heaviest and cool, respectively. There is a strong drop in temperature at the thermocline. There is no circulation of water in these three layers. If the thermocline is below the range of effective light penetration, which is quite common, the oxygen supply becomes depleted in the hypolimnion, because both photosynthesis and the surface source of oxygen are cut off. This state is known as *summer stagnation* (see Figure 7.5).

During the fall, as the air temperature drops, so does the temperature of the epilimnion until it is the same as that of the thermocline. At this point, the two layers mix. The temperature of the whole lake is now the same, and there is a complete mixing. As the temperature of the surface water reaches 4°C, it becomes more dense than water below, which is not in direct contact with the air and does not cool as rapidly at the lower levels. The denser oxygen-rich surface layer stirs up organic matter as the water sinks to the bottom; this is known as *fall turnover*.[107]

During the winter, the epilimnion, which is ice-bound, is at the lowest temperature and is thus lightest; the thermocline is at medium temperature and medium weight; and the hypolimnion is at about 4°C and heaviest. This is *winter*

---

[106]Tchobanoglous, G. and Schroeder, E. D., *Water Quality*. Reading, MA: Addison-Wesley, p. 132, 1985.
[107]Northington, D. K. and Goodin, J. R., *The Botanical World*. St. Louis: Times Mirror/Mosby College Press, p. 69, 1984.

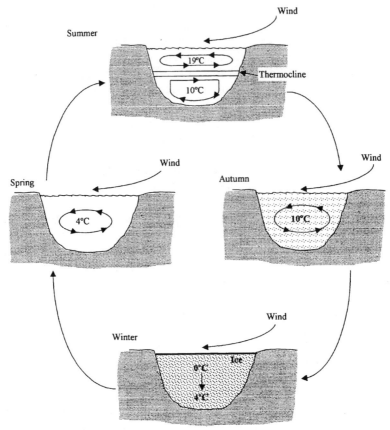

Figure 7.4 Thermal stratification of a lake. (Source: J. M. Morgan, M. D. Morgan, and J. H. Wiersma, 1986, *Introduction to Environmental Science,* Copyright ©1986 by W. H. Freeman and Company; used with permission.)

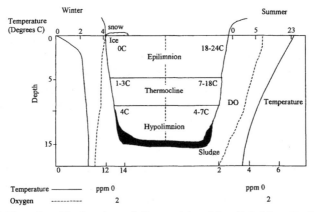

Figure 7.5 Thermal stratification of a pond. (Source: Adapted from E. P. Odum, *Fundamentals of Ecology.* Philadelphia: Saunders College Publishing, p. 310, 1971.)

stratification. In winter, the oxygen supply is usually not greatly reduced, as the low temperature solubility of oxygen is higher, and bacterial decomposition along with other life activities are operating at a low rate. When there is too much ice with heavy snow accumulation, light penetration is reduced. This reduces the rate of photosynthesis, which, in turn, causes oxygen depletion in hypolimnion, resulting in *winter kill* of fish.

In spring, as ice of the epilimnion melts (often aided by warm rains), there is a mixing of the top two layers, and as it reaches 4°C, it sinks to cause spring overturn. Odum describes this *spring overturn* phenomenon as being analogous to the lake taking a "deep breath."[108]

7.5.2 LIGHT

Viewed physically, light is part of the radiant energy of the electromagnetic spectrum. It is capable of doing work and of being transformed from one form into another. Light as radiant energy is transformed into potential energy by biochemical reactions, such as photosynthesis, or into heat. As the source of energy for photosynthesis, light is a very important factor for aquatic life. The rate of photosynthesis depends on the intensity of light and photoperiod (light hours/day). The amount of biomass and oxygen production corresponds to the rate of photosynthesis. There is a daily cycle of the amount of dissolved oxygen in water bodies based on the photosynthetic activity of plants. The amount of dissolved oxygen (DO) is maximum at 2 P.M. and minimum at 2 A.M.

7.5.3 WATER MOVEMENTS

Water movement or current is a very important limiting factor in lakes and streams. Water movements, such as wave action in lakes and the current in streams, mix the dissolved oxygen (DO) from the interphase of air and water into deeper layers. This increases the rate of absorption of oxygen from the atmosphere. The current also helps to keep the bottom clean by washing away settleable solids, thus creating a proper habitat for a large number of benthic species. Where stream current is strong, such as in riffles, specialized organisms become firmly attached to the bottom. For example, the caddis fly larvae attaches itself to the bottom substrate. In the case of fish, trout and other varieties that can swim against the current may also occupy the rapids zone of streams. Due to current flow, streams and rivers seldom have a complete depletion of dissolved oxygen (DO) in spite of organic pollution, whereas lakes and ponds can go anaerobic.

---

[108]Odum, E. P., *Fundamentals of Ecology.* Philadelphia: Saunders College Publishing, p. 311, 1971.

7.5.4 TURBIDITY

The waters of a lake or stream are often transparent. The rate of penetration of light is affected inversely by the amount of turbidity in the water. *Turbidity*, or degree of clarity, is caused by the suspended particles that block the passage of light; it often fluctuates (in surface waters) with the amount of precipitation. Turbidity, therefore, affects photosynthesis and thus lowers the number of organisms by reducing productivity. Turbidity also causes the growth of slime on the body surface of aquatic organisms, which damages the respiratory organs, such as gills, in the case of fish.

Turbidity is usually measured as *NTU*. NTU's are Nephelometric Turbidity Units, as measured with a nephelometric turbidimeter. In fieldwork, it is more common to measure turbidity by using a Secchi disk. The Secchi disk is a white disk (sometimes checkered black and white) lowered into the water column until it just disappears from view (see Figure 7.6). The depth of visual disappearance becomes the Secchi disk transparency light extinction coefficient, which will range from a few centimeters in very turbid water to 35 meters in a very clear lake such as Lake Tahoe.

7.5.5 DISSOLVED RESPIRATORY GASES

Oxygen and carbon dioxide are often limiting factors in freshwater ecosys-

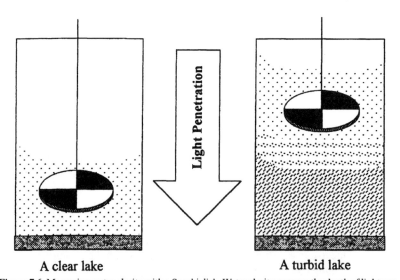

Figure 7.6 Measuring water clarity with a Secchi disk. Water clarity governs the depth of light penetration in lakes. Periodic testing with a Secchi disk may show seasonal variations in clarity. (Source: LaMotte Company; used with permission.)

tems. Carbon dioxide is produced during respiration and is essential for photosynthesis, whereas oxygen, which is produced during photosynthesis, is essential for respiration.

Several factors account for the amount of oxygen and carbon dioxide gases in water. These factors include water movements, photosynthesis, temperature and biodegradable organics. The higher the water temperature, the lower the solubility of a gas in water, and vice versa. Biodegradable organics reduce the amount of dissolved oxygen (DO) in water due to biological oxygen demand (BOD) for their composition. The amount of DO is affected inversely by the amount of carbon dioxide in the water. As oxygen is an essential ingredient for life, it becomes a very important limiting factor for aquatic life in water bodies receiving human wastes. The minimum amount of DO for normal aquatic communities is 5 mg/L. Organic pollution can dramatically reduce the amount of DO in a stream. A discussion of pollution-induced oxygen sag will be presented later.

7.5.6 BIOGENIC SALTS

Chlorides, sulfates, nitrates, phosphates of calcium, magnesium, and potassium are biogenic salts that are common, to some extent, in all freshwater ecosystems. They are essential for protoplasm synthesis. Nitrogen and phosphorous are common limiting factors in freshwater ecology. In streams and rivers, the three primary sources of basic nutrients are runoff, dissolution of rocks, and sewage discharge.

In attempting to determine limiting factors for a stream ecosystem, field observation and laboratory experimentation are necessary. Due to small variations in the natural environment, a laboratory situation is an undependable guide for determining limiting factors. Only in the field can seasonal changes in population sizes be observed.

The best way in which to determine limiting factors affecting a lake or stream is to use a combination of field and laboratory observation and experiment. An aquatic *bioassay* is one technique that can be used. In a bioassay, the scale or degree of response is determined by the rate of growth or decrease of population. Bioassays are important in evaluating water quality, because they determine the effects of liquid waste on the aquatic environments in which experimental organisms, such as fish, may be subjected to concentrations of known or suspected toxicants.

As previously stated, fish are the organisms most often studied. The species chosen should be representative of the water being studied, however. Capture of specimens taken specifically from the water source under study is recommended. Additionally, the species chosen to be studied should be the one that is most susceptible to environmental change.

The BOD test, to be discussed later, is a bioassay of the organic content of water subject to biodegradation.

7.6 LENTIC COMMUNITIES

Lakes are inland depressions containing standing water. Most lakes have outlet streams, but the lake community is quite different from the typical stream community due to the lack of current in its environment. Lentic communities are inhabited by three different classes of organisms: producers, consumers, and decomposers.

Producers are represented by rooted plants of the littoral zone and phytoplanktons of the limnetic zone. Emergent vegetation of the littoral zone consists of plants like reeds, cattails, arrowheads, and bulrushes. Floating leaf vegetation is represented by plants like the water buttercup and water lily. Submerged vegetation is formed of pondweeds and hornworts. These waterweeds have leaves that are thin and finely divided, and they provide food and resting places for clamberers.

The non-rooted plants of the limnetic zone consist of the phytoplanktons. Although these producers are microscopic and are not readily visible, they often add the distinctive green color to the water. These plants are represented by various types of algae such as blue-green algae, diatoms, filamentous green algae, and flagellate algae like *Euglena*.

Consumers in the lentic habitat are represented mainly by crustaceans, insects, mollusks, fish, annelids, helminths, rotifers, and protozoans. All five classes of the aquatic organisms (described earlier) are presented in lentic communities. *Benthos* (macrobenthos) are represented by sprawlers like crayfish, mayflies, dragonflies, clams, sludgeworms and bloodworms. *Periphytons* (microbenthos) are formed of green algae, protists, diatoms, and clamberers such as mayflies, damselflies, caddis flies, and some beetles. Making up the *nektons* are organisms such as fish, amphibians, reptiles, large crustaceans, water beetles, and water scorpions. *Neustons* are represented by water striders and whirligig beetles. *Planktons,* actually, zooplanktons, such as water fleas, copepods, rotifers and protozoans, make up the final class.

Decomposers consist of fungi and bacteria and are generally in the bottom sediments. This is not always the case, however, especially with fungi. For example, when a dead fly falls upon the surface of the water, it soon is enveloped by a halo of white fungi filaments. Fungi and bacteria work together to reduce and transform dead animals and plants into humic substances.

Although all five classes of organisms are found in lentic communities, their distribution by zone varies. For example, most of the organisms are found in the littoral zone, whereas the limnetic zone has phytoplanktons, zooplanktons, and fish. The profundal zone has some benthos varieties such as sludgeworms and annelids.

7.7 CLASSIFICATION OF LAKES

Lakes can be classified in several ways. For example, Kevern et al. classify lakes in three ways. One classification is based on productivity of the lake (or its relative richness)—the *trophic* basis of classification. A second classification is based on the times during the year that the water of a lake becomes mixed and the extent to which the water is mixed. And, a third classification is based on the fish community of lakes.[109]

For the purpose of this discussion, we use a somewhat different classification scheme than the one just described. That is, we classify lakes based on eutrophication, special types of lakes, and impoundments.

Eutrophication is a natural aging process that results in organic material being produced in abundance due to a ready supply of nutrients accumulated over time. Through natural succession, eutrophication causes a lake ecosystem to turn into a bog and, eventually, into a terrestrial ecosystem. Eutrophication has received a great amount of publicity in recent years, as humans have accelerated the eutrophication of many surface waters by the addition of wastes containing nutrients. This accelerated process is called *cultural eutrophication*. Sources of human wastes and pollution are sewage, agricultural runoff, mining, industrial wastes, urban runoff, leaching from cleared land, and landfills.

7.7.1 CLASSIFICATION OF LAKES BASED ON EUTROPHICATION

Lakes can be classified into three types based on their eutrophication stage.

(1) *Oligotrophic lakes (few foods):* they are young, deep, crystal-clear water, nutrient-poor lakes with little biomass productivity. Only a small quantity of organic matter grows in an oligotrophic lake; phytoplankton, zooplankton, attached algae, macrophytes (aquatic weeds), bacteria, and fish are present as small populations. "It's like planting corn in sandy soil, not much growth."[110] Lake Superior is an example from the Great Lakes.

(2) *Mesotrophic lakes:* it is hard to draw distinct lines between oligotrophic and eutrophic lakes, and often the term mesotrophic is used to describe a lake that falls somewhere between the two extremes. Mesotrophic lakes develop with the passage of time. Nutrients and sediments are added through runoffs, and the lake becomes more productive biologically. There is a great di-

---

[109]Kevern, N. R., King, D. L., and Ring, R., "Lake Classification Systems, Part I," *The Michigan Riparian*, p. 1, December 1999.
[110]Kevern, N. R., King, D. L., and Ring, R., "Lake Classification Systems, Part I," *The Michigan Riparian*, p. 2, December 1999.

versity of species with very low populations at first, but a shift toward higher populations with fewer species occurs. Sediments and solids contributed by runoffs and organisms make the lake shallower. At an advanced mesotrophic stage, a lake has undesirable odors and colors in certain parts. Turbidity increases, and the bottom has organic deposits. Lake Ontario has reached this stage.

(3) *Eutrophic lakes (good foods):* this is a lake with a large or excessive supply of nutrients. As the nutrients continue to be added, large algal blooms occur, fish types change from sensitive to more pollution-tolerant ones, and biomass productivity becomes very high. Populations of a small number of species become very high. The lake takes on undesirable characteristics such as offensive odors, very high turbidity, and a blackish color. This high level of turbidity can be seen in studies of Lake Washington in Seattle, Washington. Laws reports that "Secchi depth measurements made in Lake Washington from 1950 to 1979 show an almost fourfold reduction in water clarity."[111] Along with the reduction in turbidity, the lake becomes very shallow. Lake Erie is at this stage. Over a period of time, a lake eventually becomes filled with sediments as it evolves into a swamp and finally into a land area.

7.7.2 SPECIAL TYPES OF LAKES

Odum refers to several special lake types.[112]

(1) *Dystrophic (like bog lakes):* they develop from the accumulation of organic matter from outside of the lake. In this case, the watershed is often forested, and there is an input of organic acids (e.g., humic acids) from the breakdown of leaves and evergreen needles. There follows a rather complex series of events and processes, resulting finally in a lake that is usually low in pH (acid) and is often moderately clear, but color ranges from yellow to brown. Dissolved solids, nitrogen, phosphorus, and calcium are low, and humic matter is high. These lakes are sometimes void of fish fauna; other organisms are limited. When fish are present, production is usually poor. They are typified by the bog lakes of northern Michigan.

(2) *Deep ancient lakes:* these lakes contain animals found nowhere else (endemic fauna). For example, Lake Baikal in Russia.

(3) *Desert salt lakes:* these are specialized environments like the Great Salt Lake, Utah, where evaporation rates exceed precipitation rates, resulting in salt accumulation.

(4) *Volcanic lakes:* these are lakes on volcanic mountain peaks as in Japan and the Philippines.

---

[111] Laws, E. A., *Aquatic Pollution: An Introductory Text.* New York: John Wiley Sons, Inc., p. 59, 1993.
[112] Odum, E. P., *Fundamentals of Ecology.* Philadelphia: Saunders College Publishing, pp. 312–313, 1971.

(5) *Chemically stratified lakes:* examples of this type of lake include Big Soda Lake in Nevada. These lakes are stratified due to different densities of water caused by dissolved chemicals. They are *meromictic*, which means partly mixed.

(6) *Polar lakes:* these are lakes in the polar regions, with surface water temperatures mostly below 4°C.

(7) *Marl[113] lakes:* these lakes are different in that they generally are very unproductive; yet, they may have summertime depletion of dissolved oxygen in the bottom waters and very shallow Secchi disk depths, particularly in the late spring and early summer. These lakes gain significant amounts of water from springs that enter at the bottom of the lake. When rainwater percolates through the surface soils of the drainage basin, the leaves, grass, and other organic materials incorporated in these soils are attacked by bacteria. These bacteria extract the oxygen dissolved in the percolating rainwater and add carbon dioxide. The resulting concentrations of carbon dioxide can get quite high, and when they interact with the water, carbonic acid is formed.

As this acid-rich water percolates through the soils, it dissolves limestone. When such groundwater enters a lake through a spring, it contains very low concentrations of dissolved oxygen and is super-saturated with carbon dioxide. The limestone that was dissolved in the water reforms very small particles of solid limestone in the lake as the excess carbon dioxide is given off from the lake to the atmosphere. These small particles of limestone are marl, and, when formed in abundance, they cause the water to appear turbid, yielding a shallow Secchi disk depth. The low dissolved oxygen in the water entering from the springs produces low dissolved oxygen concentrations at the lake bottom.

7.7.3 IMPOUNDMENTS (SHUT-INS)

Impoundments are artificial lakes made by trapping water from rivers and watersheds. They vary in their characteristics according to the region and nature of drainage. They have high turbidity and a fluctuating water level. The biomass productivity, particularly of benthos, is generally lower than that of natural lakes.[114]

7.8 MAJOR DIFFERENCES BETWEEN LOTIC AND LENTIC SYSTEMS

Following, major differences between lotic (running water) and lentic (standing water) are highlighted.

---

[113]Kevern, N. R., King, D. L., and Ring, R., "Lake Classification Systems, Part I," *The Michigan Riparian*, pp. 4–5, December 1999.

[114]Odum, E. P., *Fundamentals of Ecology*, Philadelphia: Saunders College Publishing, p. 314, 1971.

- current
- open system (lotic) vs. closed system (lentic)
- temperature and oxygen stratification (lentic)
- bottom (substrate) types
  —lotic substrate generally more coarse due to current
  —lentic substrate generally finer due to deposition
- plankton community is an important biological component of lentic systems; usually minor in lotic
- filter feeders are an important component of lotic systems; usually minor in lentic

7.9 SUMMARY OF KEY TERMS

- *Limnology*—is the study of freshwater ecology, which is divided into two classes: lentic and lotic.
- *Lentic class (calm zone)*—consists of lakes, ponds, and swamps. This class is composed of four zones: littoral, limnetic, profundal, and benthic.
  —*Littoral zone* is the outermost shallow region of the lentic class, which has light penetration to the bottom.
  —*Limnetic zone* is the open water zone of the lentic class to a depth of effective light penetration.
  —*Euphotic* refers to all lighted regions (light penetration) formed of the littoral and limnetic zones.
  —*Profundal zone* is a deep water region beyond light penetration of the lentic class.
  —*Benthic zone* is the bottom region of a lake.
- *Lotic (washed) class*—consists of rivers and streams and is composed of two zones: rapids and pools.
  —In the *rapids zone,* the stream velocity prevents sedimentation, with a firm bottom provided for organisms specifically adapted to live attached to the substrate.
  —The *pool area* is a deeper region with a slow enough velocity to allow sedimentation. The bottom is soft due to silts and settleable solids that cause lowered DO due to decomposition.
- *Aquatic organisms*—are classified according to their mode of life. There are five classes: benthos, periphytons, plankton, pelagic (nektons), and neustons.
  —*Benthos (mud dwellers)* live within or on bottom sediments.
  —*Periphytons (Aufwucks)* live attached to plants or rocks.
  —*Planktons (drifters)* are small microscopic plants (phytoplankton) and animals (zooplankton) that move about freely with the current.

—*Pelagic (nektons)* are organisms swimming freely.
  —*Neustons* are organisms that live on the surface of the water.
- *Limiting factors*—such as temperature, light, turbidity, dissolved gases, biogenic salts, and water movements, limit the existence, growth, abundance, or distribution of organisms in natural waters.
- *Stratification*—can produce temperature-caused density differences, especially in lakes.
- *Stratified lake*—can be divided into three horizontal layers: epilimnion (upper, usually oxygenated layer); mesolimnion or hermocline (middle layer of rapidly changing temperature); and hypolimnion (lowest layer, which is subject to deoxygenation).
- *Photosynthetic rate*—depends on the intensity of light and the photoperiod.
- *High turbidity*—can reduce light penetration that can limit photosynthesis.
- *Light*—is the source of energy in the aquatic system.
- *Nutrients*—in natural waters are usually in the form of biogenic salts, which are essential for the synthesis of protoplasm.
- *Water movements*—mix oxygen into the water, distribute nutrients, and affect the type of bottom.
- *Runoff, dissolution of rocks, and sewage discharge* are the three primary sources of basic nutrients in streams and rivers.

7.10 CHAPTER REVIEW QUESTIONS

7.1 Explain the "balanced aquarium" concept.

7.2 The major difference between land and freshwater habitats is in the _____ in which they both exist.

7.3 Atmospheric air contains at least _____ times more oxygen than does water.

7.4 Limnology divides freshwater ecosystems into two groups or classes: _____ and _____ habitats.

7.5 A sessile pond is a _____ pond.

7.6 The topmost zone near the shores of a lake is known as the _____ zone.

7.7 _____ are small organisms that can feed and reproduce on their own and serve as food for small chains.

7.8 The zone of a lake not penetrated by light is called the _____ zone.

7.9 The _____ zone supports scavengers and decomposers.

7.10 Name the three zones of a lotic habitat.

7.11 _____ float or rest on the surface of water.

7.12 Define limiting factor.

7.13 List at least twelve factors that affect biological communities in streams.

7.14 What causes winter kill of fish?

7.15 Oxygen and _____ are often limiting factors in ecosystems.

7.16 The amount of DO is affected inversely by the amount of _____ in the water.

7.17 Explain eutrophication.

7.18 Lake Ontario has reached this stage.

7.19 Small particles of limestone are _____, and, when formed in abundance, they cause high turbidity in lakes.

7.20 A lotic system is _____, while a lentic system is _____.

CHAPTER 8

Stream Ecology

Fifty years ago, or even 25, the disposal of industrial wastes was not a serious problem. Individual plants were relatively small, there were few large cities. The U.S. population in 1900 was approximately one half its present level. Disposal of several forms of waste by dumping it into streams, piling in isolated areas, or discharging into the atmosphere created no major recognized problem. The pollution volume from both cities and industry with only few exceptions, was within the ability of nature to dilute to safe limits. The major problem then was sewage. Little consideration was given the treatment of water other than possibly chlorination. An engineer laid out a system which was essentially a net of collecting pipelines that conducted the liquid wastes into a river or other large body of water. Water was taken into the city from an upstream location and discharged with sewage and other waste into the downstream flow below the city. Cities and towns were so far apart that the stream could be expected to correct contamination by natural methods before the waters reached the next downstream town. Those placid days of existence have now receded into history, however.[115]

8.1 INTRODUCTION

As mentioned above, the days where natural methods in streams (e.g., the self-purification or pollution attenuation process) automatically compensate for the increase in anthropogenic (man-caused) pollution are over. To ensure that future generations will not have to deal with the devastating effects of surface water contamination, it is important for today's water/wastewater practitioners to have a basic understanding of the movement of water and the things that affect water quality and quantity.

This chapter, and the five chapters that follow—covering benthic macroinvertebrates, stream pollution, biomonitoring, self-purification, and biological sampling methods—are designed to provide the information neces-

---
[115]Grimaldi, J. V. and Simonds, R. H., *Safety Management*. Homewood, IL: Irwin, pp. 534–535, 1989.

sary to ensure a well-rounded presentation of the science of stream ecology. Additionally, this information is designed to emphasize the importance of ensuring acceptable water quality for human consumption and ecosystem survival in the future.

8.2 LIFE CYCLE OF STREAMS

As mentioned, lotic (running water) ecosystems refers primarily to the waters of streams and rivers that erode the land surface and transport and deposit materials. These running water systems are fed from precipitation that does not infiltrate the ground or evaporate. Streams, like lakes, pass through various stages. The life cycle of a stream can be divided into four stages:

(1) *Establishment of a stream:* beginning as outlets of ponds or lakes, or arising from seepage areas or springs, a stream may be a dry run or a headwater streambed before it is eroded to the level of groundwater.
(2) *Young streams:* a stream becomes permanent or youthful as its bed is eroded below the groundwater level and thus receives spring water and runoff.
(3) *Mature streams:* at this stage of development, a stream becomes wider, deeper, and turbid with low velocity. The water is generally warmer, and the bottom is formed of sand, silt, mud, or clay.
(4) *Old streams:* they have approached geologic base level. The floodplain may be very broad and flat. During the normal flow periods, the channel is refilled, and many shifting bars are developed.

An important factor to keep in mind is that streams are seasonably affected by runoff. That is, varying amounts of runoff affect the rate and volume of stream flow.

8.3 UNIQUE CHARACTERISTICS OF STREAMS

Streams have several unique characteristics. For example, temperature and dissolved oxygen (DO) are generally evenly distributed. There may be some variation, however, in DO between rapidly flowing, turbulent areas and the deeper, quiet pools of a stream due to the physical aeration of the stream. The exact amount of oxygen in aquatic systems is controlled by the solubility of gaseous oxygen in water. Because the DO in streams is usually high and evenly distributed (the amount of DO in streams and lakes is generally eight to ten parts per million), the stream organisms are adapted to this environment and have a narrow range of tolerance for DO. The exact amount of oxygen in an aquatic system is controlled by the solubility of gaseous oxygen in water. The best example of this is trout, which are adapted to high oxygen streams and cannot survive in water with DO levels below the fishkill level: below 5 mg/L.

Therefore, stream ecosystems that receive organic pollution are especially susceptible to fishkill due to the corresponding reduction in oxygen levels.

Streams exhibit a large area for land-water interchange. Most streams are primarily detritus-based food chains. This means that their primary source of energy comes not from green plants, as in most ecosystems, but from organic material from the surrounding land, which is used as food by the decomposers. Nutrients and waste products are transported by the flowing water to and away from many aquatic organisms; this process maintains a productivity that is many times greater than that in standing waters.

8.4 STREAM HABITATS[116]

Even the smallest mountain stream provides an astonishing number of different places for aquatic organisms to live (*habitats*). Streams are filled with a large variety of organisms. In a rocky stream, every rock of the substrate provides several different habitats. On the side facing upriver, organisms with special adaptations that enable them to cling to rock, do well. On the side that faces downriver, a certain degree of shelter is provided from current, but organisms are still able to hunt for food. The top of a rock, if it contacts air, is a good habitat for organisms that cannot breathe underwater and need to surface now and then. Underneath the rock is a popular habitat for organisms that hide to prevent predation.

The serious stream ecologist must know certain things about stream water and the makeup of the streambed, or substrate—or in this case, the rocky substrate and the particles that make up the substrate.

Substrate particles are measured with a metric ruler, in centimeters (cm). Because rocks can be long and narrow, we measure them twice: first the width, then the length. By adding the width to the length and dividing by two, we obtain the average size of the rock.

It is important to randomly select the rocks to be measured. Then, measure each rock. Upon completion of measurement, each rock should be classified (see Table 8.1).

Organisms that live in/on/under rocks or small spaces occupy what is known as a *microhabitat*. Some organisms make their own microhabitats: many of the caddisflies build a case about themselves and use it for their shelter (caddisflies and their "microhabitats" are discussed in greater detail in Chapter 9).

Rocks are not the only physical feature of streams where aquatic organisms can be found. For example, fallen logs and branches (commonly referred to as large woody debris, or LWD) provide excellent places for some aquatic organisms to burrow into and surfaces for others to attach themselves, as they might

---

[116]Material in this section is adapted from *The Many Habitats a Stream Provides*. cristi@ix.netcom.com, pp. 1–3, 2000.

TABLE 8.1. Stream Rock Classification Protocol.

Size (cm)	Classification
125	Boulder
25–124	Small boulder
15–24	Large cobble
5–14	Cobble
2–5	Gravel
0.1–2	Small gravel
< 0.1 and grainy	Sand
< 0.1 and floury	Silt

to a rock. They also create areas where small detritus such as leaf litter can pile up under water. These piles of leaf litter are shelter for many organisms, including large, fiercely predaceous larvae of dobsonflies (i.e., *hellgrammites;* see Chapter 9).

Another important aquatic organism habitat is found in the matter, or drift, that floats downstream. Drift is important because it is the main source of food for many fish. It may include insects such as mayflies (*Ephemeroptera*), some true flies (*Diptera*), and some stoneflies (*Plecoptera*) and caddisflies (*Trichoptera*).[117] In addition, dead or dying insects and other small organisms, terrestrial insects that fall from the trees, leaves, and other matter, are common components of drift. Among the crustaceans, amphipods (small crustaceans) and isopods (small crustaceans including sow bugs and gribbles) also have been reported in the drift.

8.5 ADAPTATIONS TO STREAM CURRENT

Current in streams is the major factor limiting the distribution of organisms. The current is determined by the steepness of the bottom gradient, the roughness of the streambed, and the depth and width of the streambed. Trautman reported that the distribution of bass and many other fish was closely correlated to stream flow rate. He found that a gradient of 7 feet/mile to 20 feet/mile was the range in which small mouth bass were found. The bass were never found at a gradient below 3 feet/mile or above 25 feet/mile. The current in streams has promoted many special adaptations by stream organisms.[118] Odum lists these adaptations as follows (see Figure 8.1):[119]

---

[117]Allen, J. D., *Stream Ecology: Structure and Function of Running Waters.* London: Chapman & Hall, pp. 221–237, 1996.
[118]Trautman, M.B., "Fish distribution and abundance correlated with stream gradients as a consideration in stocking programs." *North American Wildlife Conference,* p. 283, 1942.
[119]Odum, E.P., *Fundamentals of Ecology.* Philadelphia: Saunders College Publishing, pp. 113–114, 1971.

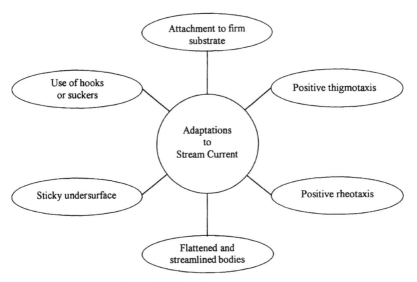

Figure 8.1 Adaptations to stream current.

(1) *Attachment to a firm substrate:* attachment is to stones, logs, leaves, and other underwater objects such as discarded tires, bottles, pipes, etc. Organisms in this group are primarily composed of the primary producer plants and animals, such as green algae, diatoms, aquatic mosses, caddisfly larvae, and freshwater sponges.

(2) *The use of hooks and suckers:* these organisms have the unusual ability to remain attached and withstand even the strongest rapids. Two diptera larvae, *Simulium* and *Blepharocera* are examples.

(3) *A sticky undersurface:* snails and flatworms are examples of organisms that are able to use their sticky undersurfaces to adhere to underwater surfaces.

(4) *Flattened and streamlined bodies:* all macroconsumers have streamlined bodies, i.e., the body is broad in front and tapers posteriorly to offer minimum resistance to the current. All nektons such as fish, amphibians, and insect larvae exhibit this adaptation. Some organisms have flattened bodies that enable them to stay under rocks and in narrow places. Examples are water penny, a beetle larva, mayfly, and stone fly nymphs.

(5) *Positive rheotaxis* (*rheo:* current; *taxis:* arrangement): an inherent behavioral trait of stream animals (especially those capable of swimming) is to orient themselves upstream and swim against the current.

(6) *Positive thigmotaxis* (*thigmo:* touch, contact): another inherent behavior pattern for many stream animals is to cling to a surface or keep the body in close contact with the surface. This is the reason that stonefly nymphs (when

removed from one environment and placed into another) will attempt to cling to just about anything, including each other.

8.6 GENERAL STREAM ADAPTATIONS

There are several basic ways for an aquatic organism to adapt to its environment.

8.6.1 TYPES OF ADAPTATIVE CHANGES[120]

Adaptative changes are classed as genotypic, phenotypic, behavioral, or ontogenetic.

(1) *Genotypic changes:* these tend to be great enough to separate closely related animals into species, such as mutations or recombination of genes. For example, salmonid has evolved a subterminal mouth (i.e., below the snout) in order to eat from the benthos.
(2) *Phenotypic changes:* these are the changes that an organism might make during its lifetime to better utilize its environment (e.g., a fish that changes sex from female to male because of an absence of males).
(3) *Behavioral changes:* these have little to do with body structure or type: a fish might spend more time under an overhang to hide from predators.
(4) *Ontogenetic changes:* these take place as an organism grows and matures (e.g., a coho salmon that inhabits streams when young and migrates to the sea when older, changing its body chemistry to allow it to tolerate saltwater).

8.6.2 SPECIFIC ADAPTATIONS

Specific adaptations observed in aquatic organisms include mouths, shape, color, aestivation, and schooling.

(1) *Mouths:* aquatic organisms such as fish change mouth shapes (morphology) depending on the food they eat. The arrangement of the jaw bones and even other head bones, the length and width of gill rakers, the number, shape, and location of teeth, and barbels change to allow fish to eat just about anything found in a stream.
(2) *Shape:* this changes to allow fish to do different things in the water. Some organisms have body shapes that push them down in the water, against the substrate, and allow them to hold their place against even strong current (e.g.,

---

[120]Adapted from *Ecology.* cristi@ix.netcom.com, pp. 1–10, 2000.

chubs, catfish, dace, and sculpins). Other organisms, especially predators, have evolved an arrangement and shape of fins that allow them to lurk without moving, lunging suddenly to catch their prey (e.g., bass, perch, pike, trout, and sunfish).

(3) *Color:* this may change within hours, to camouflage, or within days, or may be genetically predetermined. Fish tend to turn dark in clear water and pale in muddy water.

(4) *Aestivation:* this helps fishes survive in arid desert climates, where streams may dry up.

> ✓ *Note: Aestivation* refers to the ability of some fishes to burrow into the mud and wait out the dry period.

(5) *Schooling:* this serves as protection for many fishes, particularly those that are subject to predation.

8.7 BENTHIC LIFE: AN OVERVIEW[121]

The benthic habitat of stream environments is found in the streambed, or benthos. The streambed is comprised of various physical and organic materials where erosion and/or deposition is a continuous characteristic. Erosion and deposition may occur simultaneously and alternately at different locations in the same streambed. Where channels are exceptionally deep and taper slowly to meet the relatively flattened streambed, habitats may form on the slopes of the channel. These habitats are referred to as littoral habitats. Shallow channels may dry up periodically in accordance with weather changes. The streambed is then exposed to open air and may take on the characteristics of a wetland.

Silt and organic materials settle and accumulate in the streambed of slowly flowing streams. These materials decay and become the primary food resource for the invertebrates inhabiting the streambed. Productivity in this habitat depends upon the breakdown of these organic materials by herbivores. Not all organic materials are used by bottom-dwelling organisms; a substantial amount becomes part of the streambed in the form of peat.

In faster-moving streams, organic materials do not accumulate so easily. Primary production occurs in a different type of habitat found in the riffle regions where there are shoals and rocky regions for organisms to adhere to. Therefore, plants that can root themselves into the streambed dominate these regions. By plants, we are referring mostly to forms of algae, often microscopic

---

[121] From USEPA, *Lotic Ecosystems of the Streambed, ECOVIEW.* Washington, DC: United States Environmental Protection Agency, pp. 1–2, 1999.

and filamentous, that can cover rocks and debris that have settled into the streambed during summer months.

✓ *Note:* The green, slippery slime on the rocks in the streambed is representative of this type of algae.

Although the filamentous algae seems well anchored, strong currents can easily lift it from the streambed and carry it downstream where it becomes a food resource for low-level consumers. One factor that greatly influences the productivity of a stream is the width of the channel; a direct relationship exists between stream width and richness of bottom organisms. Bottom-dwelling organisms are very important to the ecosystem as they provide food for other, larger benthic organisms through consuming detritus.

8.7.1 BENTHIC PLANTS AND ANIMALS

Vegetation is not common in the streambeds of slow-moving streams; however, they may anchor themselves along the banks. Algae (mainly green and blue-green) as well as common types of water moss attach themselves to rocks in fast-moving streams. Mosses and liverworts often climb the sides of the channel onto the banks as well. Some plants similar to the reeds of wetlands with long stems and narrow leaves are able to maintain roots and withstand the current.

Slow-moving streams are dominated by aquatic insects and invertebrates. Most aquatic insects are in their larval and nymph forms, such as the blackfly, caddisfly, and stonefly. Adult water beetles and waterbugs are also abundant. Insect larvae and nymphs provide the primary food source for many fish species, including American eel and brown bullhead catfish. Representatives of crustaceans, rotifers, and nematodes (flatworms) are sometimes present. Leech, worm, and mollusk (especially freshwater mussels) abundance varies with stream conditions, but generally favors low phosphate conditions. Larger animals found in slow-moving streams and rivers include newts, tadpoles, and frogs. The important characteristic of all life in streams is adaptability to withstand currents.

8.8 SUMMARY OF KEY TERMS

- *Positive rheotaxis*—is an inherent behavioral trait of stream animals to orient themselves upstream and swim against the current.
- *Positive thigmotaxis*—is an inherent behavioral pattern of many stream animals to cling to a surface or keep the body in close contact with the surface.

- *Aestivation*—refers to the ability of some fishes to burrow into the mud and wait out the dry period.

8.9 CHAPTER REVIEW QUESTIONS

8.1 List the four stages of the stream cycle.

8.2 The exact amount of oxygen in aquatic systems is controlled by the _____ of gaseous _____ in water.

8.3 The amount of DO in streams and lakes is generally _____ to _____ parts per million.

8.4 Most streams are primarily _____ food chains.

8.5 Explain how rocks in the stream substrate are measured.

8.6 Matter that floats downstream is called _____.

8.7 _____ in streams is the major factor limiting the distribution of organisms.

8.8 _____ is an inherent behavioral trait of stream animals that allows them to orient themselves upstream and swim against the current.

8.9 List the types of adaptive changes common in streams.

8.10 In faster-moving streams, organic materials do not _____ easily.

8.11 One factor that greatly influences the productivity of a stream is the _____ of a channel.

8.12 Slow-moving streams are dominated by _____ and _____.

CHAPTER 9

Benthic Macroinvertebrates

The emphasis on aquatic insect studies, which have expanded exponentially in the last three decades, has been largely ecological. This interest in aquatic insects has grown from early limnological roots and sport fishery-related investigations of the '30s and '40s, to the use of aquatic insects as indicators of water quality during the '50s and '60s. In the '70s and '80s, aquatic insects became the dominant forms used in freshwater investigations of basic ecological questions.[122]

Freshwater invertebrates are ubiquitous; even the most polluted or environmentally extreme lotic environments usually contain some representative of this diverse and ecologically important group of organisms.[123]

9.1 BENTHIC MACROINVERTEBRATES

BENTHIC macroinvertebrates are aquatic organisms without backbones that spend at least a part of their life cycle on the stream bottom. Examples include aquatic insects, such as stoneflies, mayflies, caddisflies, midges, and beetles, as well as crayfish, worms, clams, and snails. Most hatch from eggs and mature from larvae to adults. The majority of the insects spend their larval phase on the river bottom and, after a few weeks to several years, emerge as winged adults. The aquatic beetles, true bugs, and other groups remain in the water as adults. Macroinvertebrates typically collected from the stream substrate are either aquatic larvae or adults.[124]

---

[122]Merritt, R. W. and Cummins, K. W., *An Introduction to the Aquatic Insects of North America,* 3rd ed. Dubuque, IA: Kendall/Hunt, p. 1, 1996.
[123]Hauer, F. R. and Resh, V. H., "Benthic macroinvertebrates." In *Methods of Stream Ecology.* Hauer, F. R. and Lamberti, G. A. (eds.). San Diego: Academic Press, p. 339, 1996.
[124]Dates, G., *Monitoring Macroinvertebrates.* Presented at USEPA sponsored Fifth National Volunteer Monitoring Conference Promoting Watershed Stewardship. Madison, WI: University of Wisconsin, pp. 1–36, August 3–7, 1996.

9.2 BENTHIC MACROINVERTEBRATES: INDICATOR ORGANISMS

In practice, stream ecologists observe indicator organisms and their responses (*biomonitoring*) to determine the quality of the stream environment. There are a number of methods for determining water quality based on biologic characteristics. A wide variety of indicator organisms (biotic groups) is used for biomonitoring. The most often used organisms include algae, bacteria, fish, and macroinvertebrates. A search of the database from 1993 to 1998 carried out by Vesh and Kobzina at the University of California, Berkeley, confirmed that macroinvertebrates are the most popular group.[125]

Notwithstanding their popularity, in this text, benthic macroinvertebrates are used because they offer a number of advantages:

(1) They are ubiquitous, so they are affected by perturbations in many different habitats.
(2) They are species rich, so the large number of species produces a range of responses.
(3) They are sedentary, which allows determination of the spatial extent of a perturbation.
(4) They are long-lived, which allows temporal changes in abundance and age structure to be followed.
(5) They integrate conditions temporally, so like any biotic group, they provide evidence of conditions over long periods of time.[126]

In addition, benthic macroinvertebrates are preferred as bioindicators because they are easily collected and handled by samplers; they require no special culture protocols. They are visible to the naked eye, and their characteristics are easily distinguished by samplers. They have a variety of fascinating adaptations to stream life. There are a number of excellent references available that are written in lay terms. Certain benthic macroinvertebrates have very special tolerances and thus are excellent specific indicators of water quality. Useful benthic macroinvertebrate data are easy to collect without expensive equipment. The data obtained by macroinvertebrate sampling can serve to indicate the need for additional data collection, possibly including water analysis and fish sampling.

USEPA summarizes why benthic macroinvertebrates are preferred as indicators of stream quality as follows:[127]

---

[125]Rosenberg, D. M., "A National Aquatic Ecosystem Health Program for Canada: We should go against the flow." *Bull. Entomolo. Soc. Can.*, 30(4):144–152, 1998.
[126]Rosenberg, D. M., "A National Aquatic Ecosystem Health Program for Canada: We should go against the flow." *Bull. Entomolo. Soc. Can.*, 30(4):5–6, 1998.
[127]USEPA. *Monitoring Water Quality*, Chapter 4, "Macroinvertebrates and Habitat." www.epa.gov/owow/monitoring/volunteer/stream/vms40.html, 1996.

(1) They are affected by the physical, chemical, and biological conditions of the stream.
(2) They cannot escape pollution, and they show the effects of short- and long-term pollution.
(3) They may show the cumulative impacts of pollution.
(4) They may show the impacts from habitat loss not detected by traditional ways.
(5) They are a critical part of the stream's food web.
(6) Some are very intolerant of pollution.
(7) They are relatively easy to sample and identify.

In short, the focus of this discussion is based on benthic macroinvertebrates (in regard to water quality in streams) simply because some cannot survive in polluted water, while others can survive or even thrive in polluted water. In a healthy stream, the benthic community includes a variety of pollution-sensitive macroinvertebrates. In an unhealthy stream, there may be only a few types of nonsensitive macroinvertebrates present. Thus, the presence or absence of certain benthic macroinvertebrates is an excellent indicator of stream quality.

Moreover, it may be difficult to identify stream pollution with water analysis, which can only provide information for the time of sampling (a snapshot of time). Even the presence of fish may not provide information about a polluted stream, because fish can move away to avoid polluted water and return when conditions improve. However, most benthic macroinvertebrates cannot move to avoid pollution. Thus, a macroinvertebrate sample may provide information about pollution that is not present at the time of sample collection.

Obviously, before anyone is able to use benthic macroinvertebrates to gauge water quality in a stream (or for any other reason), they must be familiar with the macroinvertebrates that are commonly used as bioindicators. Samplers need to be aware of basic insect structures before they can classify the macroinvertebrates they collect. Structures that need to be stressed include head, eyes (compound and simple), antennae, mouth (no emphasis on parts), segments, thorax, legs and leg parts, gills, abdomen, etc. Samplers also need to be familiar with insect metamorphosis—both complete and incomplete—as most of the macroinvertebrates collected are in larval or nymph stages.[128]

✓ *Note:* Providing information on basic insect structures is beyond the scope of this text. In light of this, see: *An Introduction to the Aquatic Insects of North America,* 3rd ed., Merritt, R.W. and Cummins, K.W. (eds.). Dubuque, IA: Kendall/Hunt Publishing Company, 1996.

---

[128]USEPA. *Biological Indicators of Watershed Health: Invertebrates as Indicators.* www.epa.gov/seishome/atlas/bioindicators/invertsasindicators.html, pp. 1–2, 1999; Plafkin, J. L., Barbour, M. T., Porter, K. D., Gross, S. K., and Hughes, R. M. *Rapid Bioassessment Protocols for Use in Streams and Rivers: Benthic macroinvertebrates and fish.* USEPA. Office of Water. EPA/44/4-89-001, 1989.

9.3 IDENTIFICATION OF BENTHIC MACROINVERTEBRATES

Benthic macroinvertebrates are characterized by trophic groups and mode of existence. In addition, their relationship in the food web, that is, what, or whom, they eat, is discussed.

(1) *Trophic groups:* of the trophic groups (i.e., feeding groups) that Merritt and Cummins have identified for aquatic insects, only five are likely to be found in a stream using typical collection and sorting methods:[129]

  a. *Shredders* have strong, sharp mouthparts that allow them to shred and chew coarse organic material such as leaves, algae, and rooted aquatic plants. These organisms play an important role in breaking down leaves or larger pieces of organic material to a size that can be used by other macroinvertebrates. Shredders include certain stonefly and caddisfly larvae, sowbugs, scuds, and others.

  b. *Collectors* gather the very finest suspended matter in the water by sieving the water through rows of tiny hairs. These sieves of hairs may be displayed in fans on their heads (blackfly larvae) or on their forelegs (some mayflies). Some caddisflies and midges spin nets and catch their food in them as the water flows through.

  c. *Scrapers* scrape the algae and diatoms off surfaces of rocks and debris, using their mouthparts. Many of these organisms are flattened to hold onto surfaces while feeding. Scrapers include water pennies, limpets and snails, netwinged midge larvae, certain mayfly larvae, and others.

  d. *Piercers* are herbivores that pierce plant tissues or cells and suck out the fluids. Some caddisflies do this.

  e. *Predators* eat other living creatures. Some of these are engulfers, eating their prey whole or in parts. This is very common in stoneflies and dragonflies, as well as caddisflies. Others are piercers that are like the herbivorous piercers except that they eat live animal tissues.

(2) *Mode of Existence* (habitat, locomotion, attachment, concealment):

  a. *Skaters* are adapted for "skating" on the surface where they feed as scavengers on organisms trapped in the surface film (example: water striders).

  b. *Planktonic* organisms inhabit the open water limnetic zone of standing waters (lentic: lakes, bogs, ponds). Representatives may float and swim in the open water but usually exhibit a diurnal vertical migration pattern (example: phantom midges) or float at the surface to obtain oxygen and food, diving when alarmed (example: mosquitoes).

---

[129]From Merritt, R. W. and Cummins, K. W., *Introduction to the Aquatic Insects of North America*, 3rd ed. Dubuque, IA: Kendall/Hunt, pp. 75–76, 1996.

c. *Divers* are adapted for swimming by "rowing" with the hind legs in lentic habitats and lotic pools. Representatives come to the surface to obtain oxygen, dive and swim when feeding or alarmed, and may cling to or crawl on submerged objects such as vascular plants (example: water boatmen; predaceous diving beetle).
   d. *Swimmers* are adapted for "fish-like" swimming in lotic or lentic habitats. Individuals usually cling to submerged objects, such as rocks (lotic riffles) or vascular plants (lentic), between short bursts of swimming (example: mayflies).
   e. *Clingers* have behavioral (e.g., fixed retreat construction) and morphological (e.g., long, curved tarsal claws, dorsoventral flattening, and ventral gills arranged as a sucker) adaptations for attaching to surfaces in stream riffles and wave-swept rocky littoral zones of lakes (examples: mayflies and caddisflies).
   f. *Sprawlers* inhabit the surface of floating leaves of vascular hydrophytes or fine sediments, usually with modifications for staying on top of the substrate and maintaining the respiratory surfaces free of silt (examples: mayflies, dobsonflies, and damselflies).
   g. *Climbers* are adapted for living on vascular hydrophytes or detrital debris (e.g., overhanging branches, roots, and vegetation along streams and submerged brush in lakes) with modifications for moving vertically on stem-type surfaces (examples: dragonflies and damselflies).
   h. *Burrowers* inhabit the fine sediments of streams (pools) and lakes. Some construct discrete burrows that may have sand grain tubes extending above the surface of the substrate, or the individuals may ingest their way through the sediments (examples: mayflies and midges). Some burrow (tunnel) into plant stems, leaves, or roots (miners).

9.3.1 MACROINVERTEBRATES AND THE FOOD WEB[130]

In a stream, there are two possible sources of primary energy: in-stream photosynthesis by algae, mosses, and higher aquatic plants and imported organic matter from streamside vegetation (e.g., leaves and other parts of vegetation). Simply put, a significant portion of the food that is eaten grows in the stream, like algae, diatoms, nymphs and larvae, and fish. This food that originates from within the stream is called *autochthonous*.

Most food in a stream, however, comes from outside the stream. In small, heavily wooded streams, there is normally insufficient light to support substantial in-stream photosynthesis, so energy pathways are supported largely by imported energy. A large portion of this imported energy is provided by leaves

---

[130]Benfield, E. F., "Leaf breakdown in stream ecosystems." In *Methods in Stream Ecology*, Hauer, F. R. and Lamberti, G. A. (eds.). San Diego: Academic Press, pp. 579–590, 1996.

(see the following Example). Worms drown in floods and get washed in. Leafhoppers and caterpillars fall from trees. Adult mayflies and other insects mate above the stream, lay their eggs in it, and then die in it. All of this food from outside the stream is called *allochthonous*.

9.3.2 EXAMPLE—LEAF PROCESSING IN STREAMS

Autumn leaves entering streams are nutrition-poor because trees absorb most of the sugars and amino acids (nutrients) that were present in the green leaves.[131] Leaves falling into streams may be transported short distances but are usually caught by structures in the streambed to form leaf packs. These leaf packs are then processed in place by components of the stream communities in a series of well-documented steps (see Figure 9.1).[132]

Within 24 to 48 hours of entering a stream, many of the remaining nutrients in leaves leach into the water. After leaching, leaves are composed mostly of structural materials like non-digestible cellulose and lignin. Within a few days, fungi (especially *Hyphomycetes*), protozoa, and bacteria process the leaves by microbial processing (see Figure 9.1).[133] Two weeks later, microbial conditioning leads to structural softening of the leaf and, among some species, fragmentation. Reduction in particle size from whole leaves (coarse particulate organic matter, CPOM) to fine particulate organic matter (FPOM) is accomplished mainly through the feeding activities of a variety of aquatic invertebrates collectively known as "shredders."[134] Shredders (stoneflies, for example) help to produce fragments shredded from leaves but not ingested and fecal pellets, which reduce the particle size of organic matter. The particles are

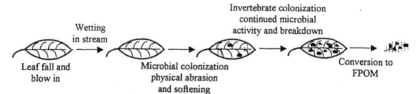

Figure 9.1 The processing or "conditioning" sequence for a medium-fast deciduous tree leaf in a temperate stream. (Source: Adapted from J. D. Allen, *Stream Ecology: Structure and Functions of Running Waters,* Tandem: Chapman & Hall, p. 114, 1996.)

---

[131]Suberkropp, K., Godschalk, G. L., and Klug, M. J. "Changes in the chemical composition of leaves during processing in a woodland stream." *Ecology,* 57:720–727, 1976; Paul, R. W., Jr., Benfield, E. F., and Cairns, J., Jr., "Effects of thermal discharge on leaf decomposition in a river ecosystem." *Verhandlungen der Internationalen Vereinigung fur Theoretische and Angewandte Limnologie,* 20:1759–1766, 1978.
[132]Peterson, R. C. and Cummins, K. W. "Leaf processing in a woodland stream," *Freshwater Biology,* 4:345–368, 1974.
[133]Barlocher, R. and Kendrick, L., "Leaf conditioning by microorganisms." *Oecologia,* 20:359–362, 1975.
[134]Cummins, K. W. "Structure and function of stream ecosystems." *BioScience,* 24:631–641, 1974; Cummins, K.W. and Klug, M. J. "Feeding ecology of stream invertebrates." *Annual Review of Ecology and Systematics,* 10:631–641, 1979.

then collected (by mayflies, for example) and serve as a food resource for a variety of micro- and macroconsumers. Collectors eat and send even smaller fragments downstream. These tiny fragments may be filtered out of the water by a true fly larva (i.e., a filterer). Leaves may also be fragmented by a combination of microbial activity and physical factors such as current and abrasion.[135]

Leaf-pack processing by all of the elements mentioned above (i.e., leaf species, microbial activity, physical and chemical features of the stream) is important. However, most important is that these integrated ecosystem processes convert whole leaves into fine particles that are then distributed downstream and used as an energy source by various consumers.

Insects that have fallen into a stream are ready-to-eat and may join leaves, exuviae, copepods, dead and dying animals, rotifers, bacteria, and dislodged algae and immature insects in their float (*drift*) downstream to be eaten.

9.4 MACROINVERTEBRATES: UNITS OF ORGANIZATION

The science of classification and naming is called *taxonomy*. Macroinvertebrates are classified and named using a *taxonomic hierarchy*. The taxonomic hierarchy for the caddisfly (a macroinvertebrate insect commonly found in streams) is shown in Table 9.1.

✓ *Note:* In the identification of macroinvertebrates, taxonomy can be at any level but should be consistent among samples. *Genus/species* provides more accurate information on ecological/environmental relationships and sensitivity to impairment. Most organisms are identified to the lowest practical level (generally genus or species) by a qualified taxonomist using a dissecting microscope. Midges (*Diptera: Chironomidae*) are mounted on slides in an appropriate medium and identified using a compound microscope. *Family* level provides a higher degree of precision among samples and taxonomists, requires less expertise to perform, and accelerates assessment results.

TABLE 9.1. Taxonomic Hierarchy: Caddisfly.

Kingdom:	Animalia (animals)
Phylum:	Arthropoda ("jointed legs")
Class:	Insecta (insect)
Order:	Trichoptera (caddisfly)
Family:	Hydropsychidae (net-spinning caddis)
Genus species:	*Hydropsyche morosa*

[135]Benfield, E. F., Jones, D. R., and Patterson, M. F. "Leaf pack processing in a pastureland stream." *Oikos*, 29:99–103, 1977; Paul, R. W., Jr., Benfield, E. R., and Cairns, J., Jr., "Effects of thermal discharge on lead decomposition in a river ecosystem." *Verhandlungen der Internationalen Vereinigung für Theoretische und Angewandte Limnologie*, 20:1759–1766, 1978.

9.5 VARIATION IN DIVERSITY OF BENTHIC MACROINVERTEBRATE SPECIES

The seasonal and spatial variations in diversity of benthic macroinvertebrate species correspond closely to *Thienemann's Principles:*[136]

(1) The greater the diversity of conditions in a locality, the larger the number of species that make up the community.
(2) The more the conditions deviate from normal, hence from the normal optima of most species, the smaller is the number of species that occur there and the greater the number of individuals of each species that do occur.
(3) The longer a locality has been in the same condition, the richer is its biotic community, and the more stable it is.

9.6 TYPICAL STREAM BENTHIC MACROINVERTEBRATES

The macroinvertebrates are the best studied and most diverse animals in streams. While it is true that non-insect macroinvertebrates, such as Oligochaeta (worms), Hirudinea (leeches), and Acari (water mites), are frequently encountered groups in lotic environments, it is the insects that are among the most conspicuous inhabitants of streams. In most cases, it is the larval stages of these insects that are aquatic, whereas the adults are terrestrial. Typically, the larval stage is extended, while the adult lifespan is short. Lotic insects are found among many different orders, and brief accounts of their biology are presented in the following sections.

9.6.1 INSECT MACROINVERTEBRATES[137]

The most important insect groups in streams are Ephemeroptera (mayflies, Section 9.6.1.1), Plecoptera (stoneflies, Section 9.6.1.2), Trichoptera (caddisflies, Section 9.6.1.3), Diptera (true flies, Section 9.6.1.4), Coleoptera (beetles, Section 9.6.1.5), Hemiptera (bugs, Section 9.6.1.6), Megaloptera (alderflies and dobsonflies, Section 9.6.1.7), and Odonata (dragonflies and damselflies, Section 9.6.1.8). The identification of these different orders is usually easy, and there are many keys and specialized references (e.g., B. W. Merritt and K. W. Cummins (eds.), *An Introduction to the Aquatic Insects of North America,* 3rd ed., Dubuque, IA: Kendall/Hunt Publishing Company,

---

[136]Mackie, G. L., *Applied Aquatic Ecosystem Concepts.* Ontario, Canada: University of Guelph, pp. 1–12, 1998.
[137]Adapted from Giller, P. S. and Malmqvist, B., *The Biology of Streams and Rivers.* Oxford, UK: Oxford University Press, pp. 83–97, 1998.

1996) available to help in the identification of species. In contrast, some genera and species, particularly the Diptera, can often only be diagnosed by specialist taxonomists. Table 9.2 provides a simple key to the main insect groups in running water systems.

As mentioned, insect macroinvertebrates are ubiquitous in streams and are often represented by many species. The size of selected orders with respect to their number of species is indicated in Table 9.3. Although the numbers refer to aquatic species, a majority are to be found in streams.

9.6.1.1 Mayflies (Order: Ephemeroptera)

Streams and rivers are generally inhabited by many species of mayflies and, in fact, most species are restricted to streams. For the experienced stream ecologist who looks upon a mayfly nymph, recognition is obtained through trained observation: abdomen with leaf-like or feather-like gills, legs with a single tarsal claw, generally (but not always) with three cerci (three "tails"; two cerci, and between them usually a terminal filament; see Figure 9.2). The experienced stream ecologist knows that mayflies are hemimetabolous insects (i.e., where larvae or nymphs resemble wingless adults) that go through many

TABLE 9.2. Simplified Key for Dominant Lotic Insects.

Adults (wings present)
a. Mouthparts forming beak on underside of head: bugs, water striders, water crickets, back swimmers, lesser water boatmen
b. Head with beak, front wings hardened, lack of veins: beetles, riffle beetles, water penny, diving beetles, whirligigs

Larvae (wings absent)
a. Wing pads (developing wings) present
 —Mouthparts form beak: bugs
 —Leaf-like gills on abdomen, tarsus 1 segmented with one claw, usually three "tails": mayflies
 —No gills on abdomen, tarsus 3 segmented with two claws, two "tails": stoneflies
 —Hinge-like lower lip below head: dragonflies and damselflies
b. Wing pads absent
 —Jointed thoracic legs present
  • prolegs present on last abdominal segment, long lateral filaments absent: caddisflies
  • prolegs present (dobsonflies) or single "tail": alderflies
  • prolegs absent from abdomen, no single "tail": beetles
 —Jointed thoracic legs absent: true flies
  • obvious head capsule: midges, blackflies
  • no obvious head capsule: craneflies

Source: Adapted from Giller, P. S. and Malmqvist, B., *The Biology of Streams and Rivers*. Oxford, UK: Oxford University Press, p. 84, 1998.

TABLE 9.3. Aquatic Insect Species (North America).

Species	N. America (North of Mexico)
Mayflies	700
Stoneflies	500
Damselflies, dragonflies	450
Water bugs	400
Fishflies, dobsonflies, alderflies	50
Spongillaflies	6
Caddisflies	1200
Beetles	1000
Aquatic moths	50
Mmidges, mosquitoes, gnats, and flies	3500*

*Includes semiaquatic species.
Note: Exact figures are subjected to constant changes due to new discoveries and revised taxonomies.
Source: Adapted from McCafferty, W. P., *Aquatic Entomology, the Fishermen's and Ecologist's Illustrated Guide to Insects and Their Relatives.* Boston: Jones and Bartlett Publishers, p. 448, 1981.

postembryonic molts, often in the range between 20 and 30. For some species, body length increases about 15% for each instar.[138]

Mayfly nymphs may live for two months to two years or more depending on the species, and range in size from 3–36 mm or more. They are mainly grazers or collector-gatherers that feed on algae and fine detritus, although a few genera are predatory. Some members filter particles from the water using hair-fringed legs or maxillary palps. Shredders are rare among mayflies. In general, mayfly nymphs tend to live mostly in unpolluted streams, where with densities of up to 10,000/square meter, they contribute substantially to secondary producers.

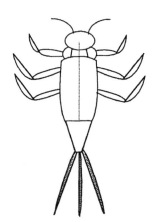

Figure 9.2 Mayfly (Order: Ephemeroptera).

[138]Humpesch, U. H., "Effect of temperature on larval growth of *Ecdyonurus dispar* (Ephemeroptera: Heptageniidae) from two English lakes." *Freshwater Biology,* 11, 441–457, 1981.

Adult mayflies resemble nymphs but usually possess two pairs of long, lacy wings folded upright; adults usually have only two cerci. The adult life span is short, ranging from a few hours to a few days, rarely up to two weeks, and the adults do not feed. Mayflies are unique among insects in that they have two winged stages, the subimago and the imago. The emergence of adults tends to be synchronous, a survival trait that possibly evolved to swamp their predators with sheer numbers, thus ensuring the survival of enough adults to continue the species.[139]

✓ *Note:* Mayflies constitute the primary basis for the sport and technique of fly-fishing and fly-tying.[140] Fishermen, in timing their trips to a stream and in choosing their lures, are attempting to imitate and take part in the feeding frenzy instigated by the mayfly reproductive cycle.

Mayflies are an important source of food for fish and other aquatic wildlife and are critical in their niche in the food chain. They are "ecological indicators" of good water quality. Thus, stream ecologists keep a close eye on mayflies. But the timeless simple elegance of their brief life cycle catches the imagination of poets, as well. The California poet Darren "Gav" Bleuel had this to say about the insects:

> *The mayfly never sees the dawn*
> *But once before his end.*
> *To think he's born*
> *upon the morn*
> *Yet not see one again.*[141]

Another example, far older:

> *When in the spring the Tisza blossoms,*
> *Thousands of mayflies play above its running ripples.*
> *None of them lives until I count hundred:*
> *The Tisza turns to a grave-yard when it blossoms.*
> *Our love was like the life of a mayfly.*
> *It ended before it really began to blossom.*
> *But with tears in my eyes I am longing*
> *For the Tisza when it blossoms!*
> —Hungarian folksong[142]

---

[139]Sweeney, B. W. and Vannote, R. L., "Population synchrony in mayflies: a predator satiation hypothesis." *Evolution,* 36, 810–821, 1982.
[140]McCafferty, W. P., *Aquatic Entomology: The Fishermen's and Ecologist's Illustrated Guide to Insects and Their Relatives.* Boston: Jones and Bartlett Publishing, Inc., p. 91, 1981.
[141]Mayflies. *Nature.* www.thirteen.org/nature/alienempire/replicators.html, p. 1, September 1, 2000.
[142]Found in McCafferty, W. P., *Aquatic Entomology: The Fishermen's and Ecologist's Illustrated Guide to Insects and Their Relatives.* Boston: Jones and Bartlett Publishing, Inc., p. 4, 1981.

9.6.1.2 Stoneflies (Order: Plecoptera)[143]

> *They clamber over the stones, keeping hold by their strong claws, always seeking the dark side of everything. They have a sidling gait and when the stones on which they live are overturned and exposed to the light they scatter like rats and drop off the edges into the water.*
> —Ann Haven Morgan[144]

Although many stream ecologists would maintain that the stonefly is a well-studied group of insects, this is not exactly the case. Despite their importance, less than 5–10% of stonefly species are well known with respect to life history, trophic interactions, growth, development, spatial distribution, and nymphal behavior.[145]

✓ *Note:* The stonefly's name undoubtedly is derived from the fact that individuals of many common species are found crawling or hiding among stones in streams or along stream banks.

Notwithstanding our lack of extensive knowledge in regard to stoneflies, enough is known to provide a fairly accurate characterization of these aquatic insects. We know, for example, that stonefly larvae are characteristic inhabitants of cool, clean streams (i.e., most nymphs occur under stones in well-aerated streams). While they are sensitive to organic pollution, or more precisely to low oxygen concentrations accompanying organic breakdown processes, stoneflies seem rather tolerant of acidic conditions. Lack of extensive gills at least partly explains their relative intolerance of low oxygen levels.

Stoneflies are drab-colored, small- to medium-sized, 1/6 to 2-1/4 inches (4 to 60 mm), rather flattened insects. Stoneflies have long, slender, many-segmented antennae and two long narrow antenna-like structures (cerci) on the tip of the abdomen (see Figure 9.3). The cerci may be long or short. At rest, the wings are held flat over the abdomen, giving a "square-shouldered" look compared to the roof-like position of most caddisflies and vertical position of the mayflies. Stoneflies have two pairs of wings. The hind wings are slightly

---

[143] Adaptation from Nielsen, G. R., *Stoneflies*. Burlington, VT: University of Vermont Extension, pp. 1–3, 1997.
[144] In McCafferty, W. P., *Aquatic Entomology: The Fishermen's and Ecologist's Illustrated Guide to Insects and Their Relatives*. Boston: Jones and Bartlett Publishing, Inc., p. 151, 1981.
[145] Stewart, K. W. and Stark, B. P., *Nymphs of North America Stonefly Genera (Plecoptera)*, Vol. 12. Denton: Thomas Say Foundation, 1988.

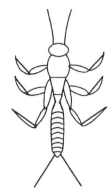

Figure 9.3 Stonefly (Order: Plecoptera).

shorter than the fore wings and much wider, having a large anal lobe that is folded fanwise when the wings are at rest. This fanlike folding of the wings gives the order its name: *"pleco"* (folded or plaited) and *"-ptera"* (wings). The aquatic nymphs are generally similar to mayfly nymphs except that they have only two cerci at the tip of the abdomen. The stoneflies have chewing mouthparts. They may be found anywhere in a nonpolluted stream where food is available. Many adults, however, do not feed and have reduced or vestigial mouthparts.

Stoneflies have a specific niche in high-quality streams where they are very important as a fish food source at specific times of the year (winter to spring, especially) and of the day. They complement other important food sources, such as caddisflies, mayflies, and midges. A more in-depth treatment of the biology and ecology of stoneflies is presented in the work of Stewart and Stark.[146]

✓ *Note:* The high water quality requirements of stonefly nymphs bar all but a very few species from habitats subject to low oxygen levels, siltation, high temperatures, and organic enrichment, and this has led to their effective use as biological indicators of environmental degradation.

9.6.1.3 Caddisflies (Order: Trichoptera)[147]

Trichoptera (Greek: *trichos*, a hair; *ptera,* wing), is one of the most diverse insect orders living in the stream environment, and caddisflies have nearly a worldwide distribution (the exception: Antarctica). Caddisflies may be catego-

---

[146] Stewart, K. W. and Stark, B. P., *Nymphs of North America Stonefly Genera (Plecoptera)*, Vol. 12. Denton: Thomas Say Foundation, 1988.

[147] Adapted from Wiggins, G. B., "Trichoptera families." In *An Introduction to Aquatic Insects of North America*, 3rd. ed. Merritt, R.W. and Cummins, K.W. (eds.). Dubuque, IA: Kendall/Hunt, pp. 309–349, 1996.

rized broadly into free-living (roving and net-spinning) and case-building species.

Caddisflies are described as medium-sized insects with bristle-like and often long antennae. They have membranous hairy wings (which explains the Latin name "Trichos") that are held tent-like over the body when at rest; most are weak fliers. They have greatly reduced mouthparts and five tarsi. The larvae are mostly caterpillar-like and have a strongly sclerotized (hardened) head with very short antennae and biting mouthparts. They have well-developed legs with a single tarsi. The abdomen is usually 10-segmented; in case-bearing species, the first segment bears three papillae, one dorsally and the other two laterally, which help hold the insect centrally in its case, allowing a good flow of water past the cuticle and gills, and the last or anal segment bears a pair of grappling hooks.

In addition to being aquatic insects, caddisflies are superb architects. Most caddisfly larvae (see Figure 9.4) live in self-designed, self-built houses, called cases. They spin out silk, and either live in silk nets or use the silk to stick together bits of whatever is lying on the stream bottom. These houses are so specialized that you can almost always identify a caddisfly larva to genus if you can see its house (case), even though there are nearly 1400 species of caddisfly in North America (north of Mexico).

Caddisflies are closely related to butterflies and moths (Order: Lepidoptera). They live in most stream habitats, and that is why they are so diverse (have so many species). Each species has special adaptations that allow it to live in the environment in which it is found.

Mostly herbivorous, most caddisflies feed on decaying plant tissue and algae. Their favorite algae is diatoms, which they scrape off of rocks. Some of them, though, are predacious.

Caddisfly larvae can take a year or two to change into adults. They then change into *pupae* (the inactive stage in the metamorphosis of many insects, following the larval stage and preceding the adult form) while still inside their cases for their metamorphosis. It is interesting to note that caddisflies, unlike stoneflies and mayflies, go through a "complete" metamorphosis.

Caddisflies remain as pupae for 2–3 weeks, then emerge as adults. When they leave their pupae, splitting their case, they must swim to the surface of the water to escape it. The winged adults fly evening and night, and some are known to feed on plant nectar. Most of them will live less than a month: like many other winged stream insects, their adult lives are brief compared to the time they spend in the water as larvae.

Caddisflies are sometimes grouped by the kinds of cases they make into five main groups: free-living forms that do not make cases, saddle-case makers, purse-case makers, net-spinners and retreat-makers, and tube-case makers.

They demonstrate their "architectural" talents in the cases they design and make. For example, a caddisfly might make a perfect, four-sided box case of

Figure 9.4 Caddis larvae, *Hydropsyche spp.*

bits of leaves and bark or tiny bits of twigs. It may make a clumsy dome of large pebbles. Others make rounded tubes out of twigs or very small pebbles. We have even found many caddisfly cases constructed of silk, emitted through an opening at the tip of the labium, used together with bits of ordinary rock mixed with sparkling quartz and red garnet, green peridot, and bright fool's gold.

Besides providing protection, caddisfly cases actually help them breathe. They move their bodies up and down, back and forth inside their cases, making a current that brings them fresh oxygen. The less oxygen there is in the water, the faster they have to move. It has been seen that caddisflies inside their cases get more oxygen than those outside of their cases, and this is why stream ecologists think that caddisflies can often be found even in still waters, where dissolved oxygen is low, in contrast to stoneflies and mayflies.

9.6.1.4 True Flies (Order: Diptera)

True or two-(*Di-*)winged (*ptera*) flies not only include the flies that we are most familiar with, like fruitflies and houseflies, they also include midges (see Figure 9.5), mosquitoes, craneflies (see Figure 9.6), and others. Houseflies and fruitflies live only on land. Some flies, however, spend nearly their whole lives in water, contributing to the ecology of streams.

True flies are in the order Diptera, one of the most diverse orders of the class Insecta, with about 120,000 species worldwide. Dipteran larvae occur almost everywhere except Antarctica and deserts where there is no running water. They may live in a variety of places within a stream: buried in sediments, attached to rocks, beneath stones, in saturated wood or moss, or in silken tubes, attached to the stream bottom. Some even live below the stream bottom.

True fly larvae may eat almost anything, depending on their species. Those

Figure 9.5 Midge larvae.

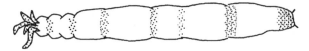

Figure 9.6 Cranefly larvae.

with brushes on their heads use them to strain food out of the water that passes through. Others may eat algae, detritus, plants, and even other fly larvae.

The longest part of the true fly's life cycle, like that of mayflies, stoneflies, and caddisflies, is the larval stage. It may remain an underwater larva anywhere from a few hours to five years. The colder the environment, the longer it takes to mature. It pupates and emerges, then, and becomes a winged adult. The adult may live four months or only a few days. While reproducing, it will often eat plant nectar for the energy it needs to make its eggs. Mating sometimes takes place in aerial swarms. The eggs are deposited back in the stream; some females will crawl along the stream bottom, losing their wings, to search for the perfect place to put their eggs. Once they lay them, they die.

Diptera serve an important role in cleaning water and breaking down decaying material, and they are a vital food source (i.e., they play pivotal roles in the processing of food energy) for many of the animals living in and around streams. However, the true flies most familiar to us are the midges, mosquitoes, and craneflies. Some midge flies and mosquitoes bite; the cranefly, however, does not bite but looks like a giant mosquito.

Like mayflies, stoneflies, and caddisflies, true flies are mostly in larval form. Like caddisflies, you can also find their pupae, because they are holometabolous insects (go through complete metamorphosis). Most of them are free-living; that is, they can travel around. Although none of the true fly larvae have the six jointed legs as seen on the other insects in the stream, they sometimes have prolegs to move around with.

Others may move somewhat like worms do, and some, the ones living in waterfalls and rapids, have a row of six suction discs that they use to move much like a caterpillar does. Many use silk pads and hooks at the ends of their abdomens to hold them fast to smooth rock surfaces.

9.6.1.5 Beetles (Order: Coleoptera)[148]

Of the more than one million described species of insect, at least one-third are beetles, making the Coleoptera not only the largest order of insects but also the most diverse order of living organisms. Even though the most diverse order

---

[148] Adapted from Hutchinson, G. E., "Thoughts on aquatic insects." *Bioscience,* 31, 495–500, 1981; Narf, R., "Midges, bugs, whirligigs and others: The distribution of insects in Lake 'U-Name-It'." *Lakeline N. Am. Lake Manage. Soc.,* 16–17, 57–62, 1997; Wetzel, R. G., *Limnology,* 2nd ed. Philadelphia: Saunders College Publishing, 1–1767, 1983.

of terrestrial insects, surprisingly, their diversity is not so apparent in running waters. Coleoptera belongs to the infraclass Neoptera, division Endpterygota. Members of this order have an anterior pair of wings (the *elytra*) that are hard and leathery and not used in flight; the membranous hindwings, which are used for flight, are concealed under the elytra when the organisms are at rest. Only 10% of the 350,000 described species of beetles are aquatic.

✓ *Note:* Some adult beetles species leave the water to seek overwintering sites or to aestivate in soil during dry periods. Overwintering adults of most species in colder regions undergo dormancy.[149]

Beetles are holometabolous. Eggs of aquatic coleopterans hatch in one or two weeks, with diapause occurring rarely. Larvae undergo from three to eight molts. The pupal phase of all coleopterans is technically terrestrial, making this life stage of beetles the only one that has not successfully invaded the aquatic habitat. A few species have diapausing prepupae, but most complete transformation to adults in two to three weeks. Terrestrial adults of aquatic beetles are typically short-lived and sometimes nonfeeding, like those of the other orders of aquatic insects. The larvae of Coleoptera are morphologically and behaviorally different from the adults, and their diversity is high.

Aquatic species occur in two major suborders, the Adephaga and the Polyphaga. Both larvae and adults of six beetle families are aquatic: Dytiscidae (predaceous diving beetles), Elmidae (riffle beetles), Gyrinidae (whirligig beetles), Halipidae (crawling water beetles), Hydrophilidae (water scavenger beetles), and Noteridae (burrowing water beetles). Five families, Chrysomelidae (leaf beetles), Limnichidae (marsh-loving beetles), Psephenidae (water pennies), Ptilodactylidae (toe-winged beetles), and Scirtidae (marsh beetles) have aquatic larvae and terrestrial adults, as do most of the other orders of aquatic insects; adult limnichids, however, readily submerge when disturbed. Three families have species that are terrestrial as larvae and aquatic as adults: Curculionidae (weevils), Dryopidae (long-toed water beetles), and Hydraenidae (moss beetles), a highly unusual combination among insects. [*Note:* Because they provide a greater understanding of a stream's condition (i.e., they are useful indicators of water quality), we focus our discussion on the riffle beetle, water penny, and whirligig beetle.]

9.6.1.5.1 Riffle Beetle (Family: Elmidae)

Riffle beetle larvae (most commonly found in running waters, hence the name riffle beetle) are up to 3/4″ long (see Figure 9.7). Their body is long, hard,

---

[149]McCafferty, W. P., *Aquatic Entomology: The Fishermen's and Ecologist's Illustrated Guide to Insects and Their Relatives.* Boston: Jones and Bartlett Publishers, Inc., p. 204, 1981.

Figure 9.7 Riffle beetle larvae.

stiff, and segmented. They have six long segmented legs on the upper middle section of their body; their back end has two tiny hooks and short hairs. Larvae may take three years to mature before they leave the water to form a pupa; adults return to the stream.

Riffle beetle adults are considered better indicators of water quality than larvae, because they have been subjected to water quality conditions over a longer period. They walk very slowly under the water (on stream bottom) and do not swim on the surface. They have small, oval-shaped bodies (see Figure 9.8) and are typically about 1/4" in length.

Both adults and larvae of most species feed on fine detritus with associated microorganisms that are scraped from the substrate, although others may be xylophagous, that is, wood eating (e.g., Lara, Elmidae). Predators do not seem to include riffle beetles in their diet, except perhaps for eggs, which are sometimes attacked by flatworms. Ecologically, their main role, therefore, seems to be in the breakdown of organic material indicated by their rich fecal production.[150]

✓ *Note:* Adult riffle beetles are able to fly when they emerge but lose this ability after they return to the water.

9.6.1.5.2 Water Penny (Family: Psephenidae)

The adult water penny is inconspicuous and often found clinging in a

Figure 9.8 Riffle beetle adult.

[150]Giller, P. S. and Malmqvist, B., *The Biology of Streams and Rivers.* London: Oxford: Oxford University Press, p. 94, 1998.

sucker-like fashion to the undersides of submerged rocks, where they feed on attached algae. The body is broad, slightly oval, and flat in shape, ranging from 4–6 mm (1/4″) in length. The body is covered with segmented plates and looks like a tiny round leaf (see Figure 9.9). It has six tiny, jointed legs (underneath). The color ranges from light brown to almost black.

✓ *Note:* Water pennies are usually found in moderate to swift currents but are occasionally found in well-oxygenated pools.

There are 14 water penny species in the United States. They live predominately in clean, fast-moving streams. Aquatic larvae live one year or more (they are aquatic); adults (they are terrestrial) live on land for only a few days. They scrape algae and plants from surfaces.

✓ *Note:* The water penny is the only larva that is almost round and keeps its head and legs hidden underneath.

9.6.1.5.3 Whirligig Beetle (Family: Gyrinidae)

Whirligig beetles are common inhabitants of streams and normally are found on the surface of quiet pools. Their bodies have pincher-like mouth parts. They have six segmented legs with claws on the middle of the body. Many filaments extend from the sides of the abdomen. And, they have four hooks at the end of the body and no tail (Figure 9.10).

✓ *Note:* When disturbed, whirligig beetles swim erratically or dive while emitting defensive secretions.

As larvae they are benthic predators, whereas the adults live on the water surface, attacking dead and living organisms trapped in the surface film. They occur on the surface in aggregations of up to thousands of individuals. Unlike the mating swarms of mayflies, these aggregations primarily serve to confuse

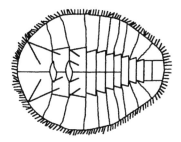

Figure 9.9 Water penny larva.

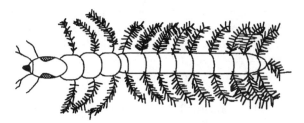

Figure 9.10 Whirligig beetle larva.

predators. Whirligig beetles have other interesting defensive adaptations. For example, the Johnston's organ at the base of the antennae enables them to echolocate using surface wave signals; their compound eyes are divided into two pairs, one above and one below the water surface, enabling them to detect aerial and aquatic predators; and they produce noxious chemicals that are highly effective at deterring predatory fish.

9.6.1.6 Water Strider ("Jesus Bugs"; Order: Hemiptera)

Water bugs are aquatic members of that group of insects called the "true bugs," most of which live on land. Moreover, unlike many other types of water insects, they do not have gills, instead getting their oxygen directly from the air. Most conspicuous and commonly known are the water striders or water skaters that ride the top of the water, with only their feet making dimples in the surface film. Like all insects, water striders have a three-part body (head, thorax, and abdomen), six jointed legs, and two antennae. They have long, dark, narrow bodies (see Figure 9.11). The underside of the body is covered with water-re-

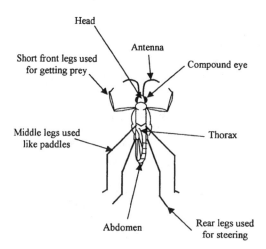

Figure 9.11 Water strider.

pellent hair. Some water striders have wings, others do not. Most water striders are over 0.2 inch (5 mm) long.

✓ *Note:* Water striders, which are also commonly known as water skaters or wheerymen, are familiar insects. In parts of Canada, they are sometimes called "Jesus bugs" because they "walk on water." Their adeptness at skating is sure to catch one's attention; in shallow waters, large round shadows are cast for each slender leg. These shadows result from the depressed but unbroken area of surface film upon which they skate.[151]

Water striders eat small insects that fall on the water's surface and larvae. Water striders are very sensitive to motion and vibrations on the water's surface, which helps them to locate prey. It pushes its mouth into its prey, paralyzes it, and sucks the insect dry. Predators of the water strider, like birds, fish, water beetles, backswimmers, dragonflies, and spiders, take advantage of the fact that water striders cannot detect motion above or below the water's surface.

✓ *Note:* Water striders in the genus *Halobates* (family Gerridae) are the only insects that are truly marine. They live on the surface of the Pacific Ocean.

9.6.1.7 Alderflies and Dobsonflies (Order: Megaloptera)[152]

> *The bright little wink under water!*
> *Mysterious wink under water!*
> *Delightful to ply*
> *The subaqueous fly*
> *And watch for the wink under water!*
> —G.E.M. Skues[153]

Larvae of all species of Megaloptera ("large wing") are aquatic and attain the largest size of all aquatic insects. Megaloptera is a medium-sized order with less than 5000 species found worldwide. Most species are terrestrial; in North America, 64 aquatic species occur.

In running waters, alderflies (Family: Sialidae) and dobsonflies (Family: Corydalidae; sometimes called hellgrammites or toe biters) are particularly important, as they are voracious predators, having large mandibles with sharp teeth.

---

[151]McCafferty, W. P., *Aquatic Entomology: The Fishermen's and Ecologist's Illustrated Guide to Insects and Their Relatives.* Boston: Jones and Bartlett Publishers, Inc., p. 182, 1981.
[152]Adapted from Evans, E. D. and Neunzig, H. H. "Megaloptera and aquatic Nueroptera." In *Aquatic Insects of North America,* 3rd ed. Merritt, R.W. and Cummins, K.W. (eds.). Dubuque, IA: Kendall/Hunt, pp. 298–308, 1996.
[153]McCafferty, W. P., *Aquatic Entomology: The Fishermen's and Ecologist's Illustrated Guide to Insects and Their Relatives.* Boston: Jones and Bartlett Publishers, Inc., p. 192, 1981.

124 BENTHIC MACROINVERTEBRATES

✓ *Note:* Because Megaloptera larvae are major predators of other aquatic insects, they form an important link in the aquatic food chain.

Alderfly brownish-colored larvae possess a single tail filament with distinct hairs. The body is thick-skinned with six to eight filaments on each side of the abdomen; gills are located near the base of each filament. Mature body size is 0.5 to 1.25 inches (see Figure 9.12). The larvae are aggressive predators, feeding on other adult aquatic macroinvertebrates (they swallow their prey without chewing); as secondary consumers, they are eaten by other, larger predators. Female alderflies deposit eggs on vegetation that overhangs water, larvae hatch and fall directly into water (i.e., into quiet but moving water). Adult alderflies are dark with long wings folded back over the body; they only live a few days.

✓ *Note:* Young alderfly larva swallow air, forming gas bubbles in their guts to allow them to float from one place in the stream to another.

Dobsonfly larvae can be rather large in size, anywhere from 25 to 90 mm (1–3″) in length. The body is stout, with eight pairs of appendages on the abdomen. Brush-like gills at the base of each appendage look like "hairy armpits" (see Figure 9.13). The elongated body has spiracles (spines) and three pairs of walking legs near the upper body and one pair of hooked legs at the rear. The head bears four segmented antennae, small compound eyes, and strong mouth parts (large chewing pinchers). Coloration varies from yellowish, brown, gray and black, to often mottled. Dobsonfly larvae, commonly known as hellgrammites, are customarily found along stream banks under and between stones. As indicated by the mouthparts, they are predators and feed on all kinds of aquatic organisms.

✓ *Note:* Fishermen use dobsonflies for bait because the trout feed on them.

9.6.1.8 Dragonflies and Damselflies (Order: Odonata)

The Odonata (dragonflies, suborder Anisoptera; and damselflies, suborder Zygoptera) is a small order of conspicuous, hemimetabolous insects (lack a pupal stage) of about 5000 named species and 23 families worldwide. Odonata is a

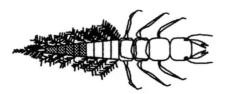

Figure 9.12 Alderfly larva.

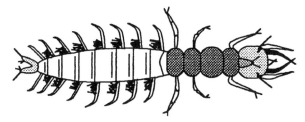

Figure 9.13 Dobsonfly larva.

Greek word meaning "toothed one." It refers to the serrated teeth located on the insect's chewing mouthparts (mandibles).

Characteristics of dragonfly and damselfly larvae include the following:

- large eyes
- three pairs of long segmented legs on the upper middle section (thorax) of the body
- large scoop-like lower lip that covers the bottom of the mouth
- no gills on sides of or underneath abdomen

✓ *Note:* Dragonflies and damselflies are unable to fold their four elongated wings back over the abdomen when at rest.

Dragonflies and damselflies are medium to large insects with two pairs of long equal-sized wings. The body is long and slender, with short antennae. Immature stages are aquatic, and development occurs in three stages (egg, nymph, adult).

✓ *Note:* Odonata are one of the few groups of aquatic insects in which nonscientific, or common, names have been liberally applied to species. This is undoubtedly due to the interest that naturalists have long had in them. Most common names are based either on characteristics of the terrestrial adults or on transliterations of the Latinized scientific names (e.g., *Gomphus olivaceus* or the Olive Clubtail; *Cordulegaste sayi* or Say's Biddie).[154]

Dragonflies are also known as darning needles. [*Note:* Myths about dragonflies warned children to keep quiet, or else the dragonfly's "darning needles" would sew the child's mouth shut.] The nymphal stage of dragonflies consists of grotesque creatures, robust and stoutly elongated. They do not have long "tails" (see Figure 9.14) and are commonly gray, greenish, or brown to black in

---

[154]McCafferty, W. P., *Aquatic Entomology: The Fishermen's and Ecologist's Illustrated Guide to Insects and Their Relatives*. Boston: Jones and Bartlett Publishers, Inc., p. 139, 1981.

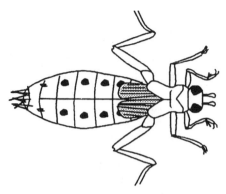

Figure 9.14 Dragonfly nymph.

color. They are medium to large aquatic insects ranging from 15 to 45 mm. Their legs are short and used for perching. They are often found on submerged vegetation and at the bottom of streams in the shallows, but are rarely found in polluted waters. Food consists of other aquatic insects, annelids, small crustacea, and mollusks. Transformation occurs when the nymph crawls out of the water, usually onto vegetation. There it splits its skin and emerges prepared for flight. The adult dragonfly is a strong flier, capable of great speed (>60 mph) and maneuverability. When at rest, the wings remain open and out to the sides of the body. A dragonfly's freely movable head has large, hemispherical eyes (nearly 30,000 facets each), which the insects use to locate prey. Dragonflies eat small insects, mainly mosquitoes (large numbers of mosquitoes), while in flight. Depending on the species, dragonflies lay hundreds of eggs by dropping them into the water and leaving them to hatch or by inserting eggs singly into a slit in the stem of a submerged plant. The incomplete metamorphosis (egg, nymph, mature nymph, adult) can take two to three years. Nymphs are often covered by algal growth.

✓ *Note:* Adult dragonflies are sometimes called "mosquito hawks" because they eat such a large number of mosquitoes that they catch while they are flying.

Damselflies are smaller and more slender than dragonflies. They have three long, oar-shaped feathery tails that are actually gills, and long slender legs (see Figure 9.15). They are gray, greenish, or brown to black in color. Their habits are similar to those of dragonfly nymphs, and they emerge from the water as adults in the same manner. The adult damselflies are slow and seem uncertain in flight. Wings are commonly black or clear, and their body is often brilliantly colored. When at rest, they perch on vegetation with their wings closed upright. Damselflies mature in one to four years. Adults live for a few weeks or months.

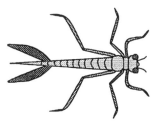

Figure 9.15 Damselfly nymph.

Unlike the dragonflies, adult damselflies rest with their wings held vertically over their backs. They mostly feed on live insect larvae.

✓ *Note:* Relatives of the dragonflies and damselflies are some of the most ancient of the flying insects. Fossils have been found of giant dragonflies with wingspans up to 720 mm that lived long before the dinosaurs.

9.6.2 NON-INSECT MACROINVERTEBRATES[155]

Three frequently encountered groups in running water systems are Oligochaeta (worms), Hirudinea (leeches), and Gastropoda (lung-breathing snails). They are by no means restricted to running water conditions, and the great majority of them occupy slow-flowing marginal habitats where the sedimentation of fine organic materials takes place. [*Note:* Non-insect macroinvertebrates are important to our discussion of stream ecology and self-purification because many of them are used as bioindicators of stream quality.]

9.6.2.1 Oligochaeta (Family Tuificidae, Genus *Tubifex*)[156]

Tubifex worms (commonly known as sludge worms) are unique in the fact that they build tubes. Sometimes, there are as many as 8000 individuals per square meter. They attach themselves within the tube and wave their posterior end in the water to circulate the water and make more oxygen available to their body surface (see Figure 9.16). These worms are commonly red, because their blood contains hemoglobin. *Tubifex* worms may be very abundant in situations when other macroinvertebrates are absent; they can survive in very low oxygen

---

[155]Adapted from Giller, P. S. and Malmqvist, B., *The Biology of Streams and Rivers*. Oxford, UK: University of Oxford Press, p. 83, 1998.
[156]Spellman, F. R., *Microbiology for Water/Wastewater Operators*. Revised ed. Lancaster, PA: Technomic Publishing Company, Inc., pp. 92–93, 2000.

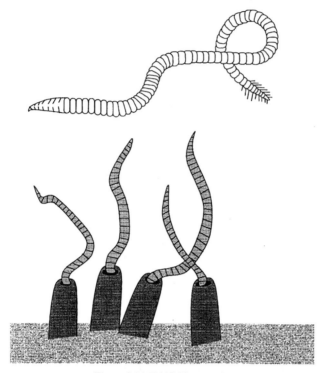

Figure 9.16 Tubificid worm(s).

levels and can live with no oxygen at all for short periods of time. They are commonly found in polluted streams, where they feed on sewage or detritus. The bottoms of severely polluted streams can literally be covered with a "writhing" mass of these worms.[157]

9.6.2.2 Hirudinea (Leeches)

There are many different families of leeches, but all have common characteristics. They are soft-bodied worm-like creatures that are flattened when extended. Their bodies are dull in color, ranging from black to brown and reddish to yellow, often with a brilliant pattern of stripes or diamonds on the upper body. Their size varies within species but generally ranges from 5 mm to 45 cm when extended. Leeches are very good swimmers, but they typically move in an inchworm fashion. They are carnivorous and feed on other organisms ranging from snails to warm-blooded animals. Leeches are found in warm, protected shallows under rocks and other debris.

[157]Pennack, R. W., *Fresh-water Invertebrates of the United States,* 3rd ed. p. 189, 1989.

9.6.2.3 Gastropoda (Lung-Breathing Snail)

Lung-breathing snails (pulmonates) may be found in streams that are fairly clean. However, their dominance may indicate that dissolved oxygen levels are low. These snails are different from right-handed snails because they do not breath under water by use of gills but instead have a lung-like sac called a pulmonary cavity that they fill with air at the surface of the water. When the snail takes in air from the surface, it makes a clicking sound. The air taken in can enable the snail to breath underwater for long periods of time, sometimes hours.

Lung-breathing snails have two characteristics that help to identify them. First, they have no operculum, or hard cover over the opening to their body cavity. Second, snails are either "right-handed" or "left-handed," and the lung-breathing snails are "left-handed." This can be determined by holding the shell so that its tip is upward and the opening faces us. If the opening is to the left of the axis of the shell, the snail is termed sinistral, left-handed. If the opening is to the right of the axis of the shell, the snail is termed dextral, right-handed, and it breathes with gills.

Snails are animals of the substrate and are often found creeping on all types of submerged surfaces in water from 10 cm to 2 m deep. Traits that determine lung-breathing snails include the following:

- no plate-like covering (operculum) over the shell opening
- has shell that spirals with opening usually on the left side (if tip is pointed upward and opening is facing you), or shell that is coiled in one plane, or shell that is domed or hat-shaped with no coils

9.7 SUMMARY OF KEY TERMS

- *Abdominal*—pertains to structures or parts of the abdomen (the third and most posterior major body region).
- *Appendage*—is an extension from the main part of an animal's body. Appendages include arms, legs, and tails. Other appendages may serve functions such as breathing or feeling, or they may simply be present for appearance.
- *Aquatic organisms*—live primarily or exclusively in water.
- *Terrestrial organisms*—live primarily or exclusively on land.
- *Body structure*—in insects is composed of three main divisions as follows:
  —Head is the front part that usually includes the mouth, jaws, and associated structures, the eyes, and the antennae.
  —Thorax is the middle part where the legs are usually attached.
  —Abdomen is the third part, and that part which is farthest from the head.

- *Gills*—are special structures developed for the purpose of removing oxygen from water to allow the insect to "breathe." Gills may be flat, finger-like, thread-like, or feather-like in shape. They may also be located on any of the three major body divisions.
- *Metamorphosis*—refers to the series of changes that an insect undergoes through its life cycle. Complete metamorphosis refers to a life cycle that includes an egg, larva, pupa, and adult. The life cycle of a butterfly is a complete metamorphosis. In contrast, some aquatic insects transform directly from a larva to an adult; a process known as incomplete metamorphosis. Many insects will have more than one larval stage and other variations of the simple, but complete, metamorphosis.
- *Larva* (pl. *larvae*)—is the first mobile life stage of an insect. Most insects hatch from eggs to a larval form. The larva may have several stages. Some insects follow the larval form with a pupal stage, others go directly to the adult.
- *Adult*—is the final stage of an insect. The primary function of many adult insects is simply to reproduce and lay eggs. Some do not even eat while in the adult life form.
- *Pupa* (pl. *pupae*)—is a transitional life stage, usually stationary or attached, that follows the larval stage and precedes the adult of many insects. Insects that spend part of their life cycle as a pupa are said to undergo complete metamorphosis. Some familiar pupae include cocoons (moths) or chrysalises (butterflies).
- *Dissolved Oxygen (DO)*—is the form of oxygen that many aquatic macroinvertebrates "breathe" to survive. Animals that live on land remove oxygen from the air they inhale. Aquatic animals must find some other source of oxygen. Many of them use gills or other organs to remove the oxygen that is dissolved in water. Water will have higher levels of oxygen if it is fast moving, especially in riffles, where air can be trapped and absorbed. As water warms, less oxygen will stay dissolved, and it will be harder for these animals to survive. Streams that have lots of trees to provide shade will stay cooler, have higher levels of dissolved oxygen, and be healthier for the animals and fish that live in them.
- *Detritivore*—is an animal that eats exclusively or primarily dead and decomposing plant materials.
- *Herbivore*—is an animal that eats exclusively or primarily living plants.
- *Predator*—is an animal that stalks, kills, and eats other living animals.
- *Proleg*—is a small leg-like appendage present on some larva and worms. These may or may not be used to help the animal move.
- *Operculum*—is a hard plate-like cover over the shell of gilled snails.
- *Elytron* (pl. *elytra*)—is the hardened and commonly plate-like modified forewing of beetle adults.
- *Palp*—is an arm-like appendage, especially of the maxilla or labium.
- *Sclerotized*—refers to hardened body wall.

9.8 CHAPTER REVIEW QUESTIONS

9.1 _____ are aquatic organisms without backbones that spend at least a part of their life cycle on the stream bottoms.

9.2 List the five trophic groups likely to be found in a stream using typical collection and sorting methods.

9.3 _____ are adapted for swimming by "rowing" with the hind legs in lentic habitats and lotic pools.

9.4 Most food in a stream comes from _____ the stream.

9.5 _____ level provides a higher degree of precision among samples and taxonomists.

9.6 List *Thienemann's Principles*.

9.7 The inactive stage in the metamorphosis of many insects, following the larval stage and preceding the adult form is the _____ stage.

9.8 _____ larvae occur almost everywhere except Antarctica and deserts where there is no running water.

9.9 _____ are small leg-like appendages present on some larva and worms.

9.10 Ecologically, their main role seems to be in the breakdown of organic material indicated by their rich faecal production: _____.

Matching Exercise: Match the definition listed in Part A with the terms listed in Part B by placing the correct letters in the blanks.

Part A:

(1) Typically collected from the stream substrate as either aquatic larvae or adults: _____
(2) Helps to determine the quality of the stream environment: _____
(3) Have strong, sharp mouthparts that allow them to shred and chew coarse organic material: _____
(4) Scrape the algae and diatoms off of surfaces of rocks and debris using their mouthparts: _____

(5) Inhabit the open water limnetic zone of standing waters: _____
(6) Usually have modifications for staying on top of the substrate: _____
(7) Provides a large portion of energy in a stream: _____
(8) Food from outside the stream: _____
(9) Fall into a stream: _____
(10) Provides a higher degree of precision among samples and taxonomists: _____
(11) Short-lived flyer: _____
(12) Casting of skin: _____
(13) Seems rather tolerant to acidic conditions: _____
(14) Net-spinners: _____
(15) Order: Lepidoptera: _____
(16) Houseflies and fruitflies: _____
(17) Have hard and leathery wings: _____
(18) Long-toed water beetles: _____
(19) Adults are better indicators of water quality: _____
(20) Almost round with its head and legs hidden from view: _____
(21) Johnston's organ: _____
(22) Walks on water: _____
(23) Large wing: _____
(24) Deposits eggs on vegetation that overhangs water: _____
(25) Trout love them: _____
(26) "Darning needles": _____

Part B:

a. ready-to-eat insects
b. shredders
c. biomonitoring
d. molt
e. water penny
f. dobsonflies
g. butterfly
h. macroinvertebrates
i. leaves
j. caddisfly
k. whirligig beetle
l. dragonflies
m. planktonic
n. riffle beetle
o. alderfly
p. sprawlers
q. mayfly
r. megaloptera
s. true flies
t. scrapers
u. stonefly
v. beetles
w. water striders
x. allochthonous
y. Dryopidae
z. family level

CHAPTER 10

Stream Pollution

"Most Endangered" U.S. Rivers[158]

American Rivers (Washington, D.C.) announced its 15th annual list of the most endangered rivers in the United States in April, noting that dams, levees, and stabilized riverbanks are destroying habitat and contributing to the extinction of native fish and wildlife across the nation.

"America's native fish are homeless in most parts of the country," says Rebecca Wodder, president of American Rivers. "We have straightened the curves, blocked the flows, and hardened the banks of thousands of miles of waterways, wiping out habitats and making it difficult for our nation's rivers to support native fish and wildlife," she says.

Scientists believe that habitat loss could contribute to the extinction of hundreds of freshwater species in the United States by the end of the 21st century.

Top 10 Most Endangered U.S. Rivers

1. Lower Snake River (Wash.)
2. Missouri River (Mont., N.D., S.D., Neb., Iowa, Kan., Mo.)
3. Ventura (Calif.)
4. Copper River (Calif.)
5. Tri-State River Basins (Ga., Ala., Fla.)
6. Coal River (W. Va.)
7. Rio Grande (Colo., N.M., Texas)
8. Mississippi and White Rivers (Minn., Wis., Ill., Iowa, Mo., Ky., Tenn., Ark., Miss., La)
9. North Fork Feather River (Calif.)
10. Clear Creek (Texas)

10.1 WHAT IS STREAM POLLUTION?[159]

PEOPLE'S opinions differ in what they consider to be a pollutant on the basis of their assessment of benefits and risks to their health and economic

---

[158]"Conservation Group Announces 'Most Endangered' U.S. Rivers." In *Water Environment & Technology* (*WE&T*), p. 13, July 2000.
[159]From Spellman, F. R., *The Science of Environmental Pollution*. Lancaster, PA: Technomic Publishing Company, Inc., pp. 4–5, 1999.

well-being. For example, visible and invisible chemicals spewed into water by an industrial facility might be harmful to people and other forms of life living nearby and in the stream itself. However, if the facility is required to install expensive pollution controls, forcing the industrial facility to shut down or to move away, workers who would lose their jobs and merchants who would lose their livelihoods might feel that the risks from polluted air and water are minor weighed against the benefits of profitable employment. The same level of pollution can also affect two people quite differently. Some forms of water pollution, for example, might cause only a slight irritation to a healthy person but cause life-threatening problems to someone with autoimmune deficiency problems. Differing priorities lead to differing perceptions of pollution (concern about the level of pesticides in foodstuffs prompting the need for wholesale banning of insecticides is unlikely to help the starving). Public perception lags behind reality because the reality is sometimes unbearable. Pollution is a judgement, and pollution demands continuous judgement.

10.2 STREAM POLLUTION LAWS[160]

Existing laws and their implementing regulations treat water narrowly, as if surface and groundwater were not connected, as if point and non-point sources of pollution could be treated in isolation. They undervalue the immense diversity of goods and services supplied by aquatic ecosystems.

The Clean Water Act (CWA) mandates, "to restore and maintain the physical, chemical, and biological integrity of the nation's waters."[161]

For 25 years, this mandate was largely ignored in water policy.[162] Three approaches to the use and management of water resources kept the focus narrow, incomplete, and inadequate:

(1) *Water was viewed as a fluid for humans to use.* Too many water resource professionals saw "the forms of life in a [stream as] purely incidental, compared with the task of a [stream], which is to conduct water runoff from an area toward ocean."[163]

(2) *Pollution was the only threat to water resources, and dilution was the solution.* People managed for "water quality" (degrees of chemical contamina-

---

[160] Adapted from Karr, J. R., *Rivers as Sentinels: Using the Biology of Rivers to Guide Landscape Management.* pacnwfin.mss, pp. 3–4, revised August 1996.
[161] USEPA. *Summary of State Biological Assessment Programs for Streams and Rivers.* Washington, DC: Environmental Protection Agency, EPA 230-R-96-07, p. 3 1996; USEPA, Biological Assessment Methods, *Biocriteria, and Biological Indicators: Bibliography of Selected Technical, Policy, and Regulatory Literature.* Washington, DC: Enviromental Protection Agency, EPA 230-B-96-001, p. 33, 1996.
[162] Karr, J. R. and Dudley, D. R., "Ecological perspective on water quality goals." *Environmental Management,* 5:55–68, 1981; USEPA, *Biological Criteria: National Program Guidances for Surface Waters.* Washington, DC: Environmental Protection Agency, EPA 440-5-90-004, 1990; Karr, C., "Biological integrity and the goal of environmental legislation: lessons for conservation biology." *Conservation Biology,* 4:66–84, 1991.
[163] Einstein, H. A., "Sedimentation (suspended solids)." In *River Ecology and Man.* Oglesby, C., Carlson, A., and McCann, J. (eds.). New York: Academic Press, pp. 309–318, 1972.

tion). In 1965, an Illinois water official observed, "Regardless of how one may feel about the discharge of waste products into surface waters, it is accepted as a universal practice and . . . a legitimate use of stream waters."[164] Surface waters existed to receive the discharge of human society.

(3) *Only a few aquatic species "counted" as being important to human society.* Society sought to maximize sport or commercial harvest of selected species. Production—larger harvests of fish or shellfish—became the goal, and technofixes like hatcheries became the means to supplement falling wild populations.[165] Fish ladders helped migrating adults pass upstream over dams, but no provisions were made for helping young fish go around the dams as they migrated downstream toward the ocean. Many biologists removed large woody debris from stream channels to make passage easier, never mind that fish had been passing such barriers for centuries, or that the wood actually created fish habitat.[166]

The first two attitudes did not give any value to the life-forms associated with stream ecosystems. Although these three philosophies have not been abandoned, a growing number of water resource professionals recognize their inadequacies.

10.2.1 TMDL RULE[167]

On July 11, 2000, a U.S. Environmental Protection Agency (USEPA) administrator signed a rule that revised the Total Maximum Daily Load (TMDL) program and made related changes to the National Pollutant Discharge Elimination System (NPDES) and Water Quality Standards programs (65 FR 43585, July 13). According to President Clinton, EPA's move was a "critical, common-sense step" to clean up the nation's waterways.

The USEPA points out that over 20,000 water bodies across America have been identified as polluted by States, Territories, and authorized Tribes. These waters include over 300,000 stream/river and shoreline miles and 5 million acres of lakes. The overwhelming majority of people in the United States live within 10 miles of one of these polluted waters.

The Clean Water Act (CWA) provides special authority for restoring polluted waters. The Act calls on states to work with interested parties to develop

---

[164]Evans, R., "Industrial wastes and water supplies." *Journal of American Water Works Association,* 57:625–628, 1965.
[165]Meffe, G. K., "Techno-arrogance and halfway technologies: salmon hatcheries on the Pacific Coast of North America." *Conservation Biology,* 6:350–354, 1992.
[166]Maser, C. and Sedell, J. R., *From the Forest to the Sea: The Ecology of Woodland Streams, Rivers, Estuaries, and Oceans.* Delray Beach, FL: St. Lucie Press, p. 37, 1994.
[167]"U.S. EPA signs TMDL Rule despite congressional protests." *Water Environment & Technology (WE&T),* p. 40, August 2000; USEPA, *Total Maximum Daily Load (TMDL) Program.* Washington, DC: U.S. Environmental Protection Agency, EPA 841-F-00-008, pp. 1–4, July 2000.

Total Maximum Daily Loads (TMDLs) for polluted waters. A TMDL is essentially a "pollution budget" designed to restore the health of the polluted body of water.

10.2.1.1 Goals of TMDL Rule

The TMDL rule will make thousands more streams/rivers, lakes, and coastal waters safe for swimming, fishing, and healthy population of fish and shellfish. Key provisions of the TMDL Rule include the following:

- It requires states to develop more detailed listing methods and comprehensive lists of polluted water bodies, which must be submitted to the USEPA every four years. The lists also may include threatened waters.
- It requires states to prioritize water bodies and develop TMDLs first for those that are drinking water sources or that support endangered species. Once a TMDL is developed, the rule requires states to establish a cleanup schedule that would enable polluted water bodies to achieve water quality standards within 10 years (within 15 years if the state requests and EPA grants an extension).
- TMDL development must include an implementation plan that identifies specific actions and schedules for meeting water quality goals and addresses point and non-point pollution sources, according to the rule. The rule also requires that runoff controls be installed five years after this plan is developed, if practicable, and that TMDL allocations for non-point sources be pollution specific, implemented expeditiously, met through effective programs, and supported by adequate water quality funding.
- The rule does not require new permits for forestry, livestock, or aquaculture operations. It also does not require "offsets" for new pollution discharges to impaired waters prior to TMDL development.

10.3 STREAM POLLUTANTS

With regard to stream pollution, no single public concern is greater than when a highly visible massive fishkill occurs in a local stream. Moreover, when local, state, and national media announce that thousands of fish have died in some particular body of water, the public clamors for remedy. Surface waters such as local streams can have a profound impact upon the public. For example, when one recognizes that the public may drink the stream water, eat the fish from the stream, and use the stream as a recreational resource, then it becomes quite apparent that the public has a stake in the quality of its local stream water. There is irony in all this, however, as apparent in the following discourse presented by Halsam:

Man's actions are determined by his expediency. If it makes man's life more convenient, less expensive or pleasanter, the stream and its aquatic life will be sacrificed. Actions to benefit the stream come only when its state displeases man: when it carries cholera or cadmium, when its ugliness offends, or when species or habitats he now thinks important are being lost.[168]

✓ *Note:* When registered voters were asked what is the most important environmental problem facing the nation (U.S.), they responded:[169]

Air pollution	26%
Unsafe drinking water	11%
Water pollution	11%
Toxic/hazardous waste	10%
Dealing with household garbage/waste	10%

Aquatic pollution in local streams is composed of storm-water runoff, wastes from industry, and wastes from homes and commercial enterprises. Several types of aquatic pollutants have caused problems in natural bodies of water. Some of the common pollutants and their effects will be discussed in this section. Miller, for convenience, breaks down biological, chemical, and physical forms of water pollution into the following eight major types (see Figure 10.1):[170]

(1) Bacteria, viruses, protozoa, and parasites—*disease-causing agents*
(2) Domestic sewage, animal manure, and other biodegradable organic wastes that deplete water of dissolved oxygen—*oxygen-demanding wastes*
(3) Acids, salts, toxic metals, and their compounds—*water-soluble inorganic chemicals*
(4) Water-soluble nitrate and phosphate salts—*inorganic plant nutrients*
(5) Insoluble and water-soluble oil, gasoline, plastics, pesticides, cleaning solvents, and many others—*organic chemicals*
(6) Insoluble particles of soil, silt, and other inorganic and organic materials that can remain suspended in water—*sediment or suspended matter*
(7) Heat
(8) Radioactive substances

Some of the common pollutants that have direct impact upon stream ecology and that are pertinent to this discussion are discussed in the following sections.

---

[168]Halsam, S. M., *River Pollution: An Ecological Perspective.* New York: Belhaven Press, p. 6, 1990.
[169]*USA Today.* "Pollution is top environment concern," August 29, 2000.
[170]Miller, G. T., *Environmental Science,* 2nd ed. Belmont, CA: Wadsworth, p. 347, 1988.

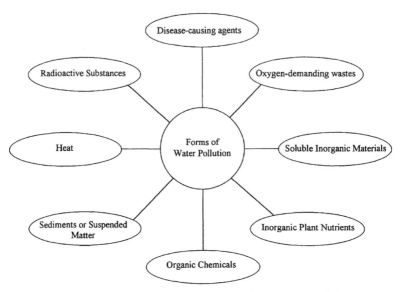

Figure 10.1 Biological, chemical, and physical forms of water pollution.

10.3.1 pH

Both acidic and alkaline wastes may be generated by mine drainage, various industrial wastes, and by acid deposition (acid rain). Streams located in rural settings are not exempt from acidification. Mason points out that several mining operations are situated in rural settings and end up discharging waste product into streams that would otherwise be normal and quite clean.[171]

A sharp change in pH in a natural stream may cause the death of most organisms in the stream. To ensure the protection of aquatic organisms, discharged wastes should not lower the pH below 6.5 or raise it above 8.5. Recent studies have shown that some organisms are capable of acclimating to alkaline waters. However, to date, there has been no evidence that organisms can acclimate to more acidic water conditions. The sensitivities of various aquatic organisms to lowered pH, based on studies conducted in Scandinavian lakes, are provided in Table 10.1.

10.3.1.1 Effects of Mine Drainage on Aquatic Macroinvertebrates[172]

According to Kimmel " . . . The influx of untreated acid mine drainage into streams can severely degrade both habitat and water quality often producing an

---

[171]Mason, C. F., "Biological aspects of freshwater pollution." In *Pollution: Causes, Effects, and Control*. Harrison, R. M. (ed.). Cambridge, Great Britain: The Royal Society of Chemistry, p. 133, 1990.
[172]Earle, J. and Callaghan, J., *Impact of Mine Drainage on Aquatic Life, Water Uses, and Man-made Structures*. Pennsylvania Dept. of Environmental Protection (PA DEP), pp. 1–13, 1998.

TABLE 10.1. Sensitivities of Aquatic Organisms to Lowered pH.

pH	
6.0	Crustaceans, mollusks, etc., disappear.
	White moss increases.
5.8	Salmon, char, trout, and roach die.
	Sensitive insects, phytoplankton, and zooplankton die.
5.5	Whitefish, grayling die.
5.0	Perch, pike die.
4.5	Eels, brook trout die.

Source: Reproduced by permission from *Pollution* by J. N. Lester, Royal Society of Chemistry, Cambridge, p. 109, 1990.

environment devoid of most aquatic life and unfit for desired uses. The severity and extent of damage depends upon a variety of factors including the frequency, volume, and chemistry of the drainage, and the size and buffering capacity of the receiving stream."[173]

Mine drainage is a toxic cocktail of intricately mixed elements that interact to cause a variety of effects on stream life that are difficult to separate into individual components. Toxicity is dependent on discharge volume, pH, total acidity, and concentration of dissolved metals. pH is the most critical component, because the lower the pH, the more severe the potential effects of mine drainage on aquatic life. The overall effect of mine drainage is also dependent on the dilution rate of flow, pH, and alkalinity or buffering capacity of the receiving stream. The higher the concentration of bicarbonate and carbonate ions in the receiving stream, the higher the buffering capacity and the greater the protection of aquatic life from adverse effects of acid mine drainage.[174] Alkaline mine drainage with low concentrations of metals may have little discernible effect on receiving streams. Acid mine drainage with elevated metals concentrations discharging into headwater streams or lightly buffered streams can have a devastating effect on aquatic life. Secondary effects such as increased carbon dioxide tensions, oxygen reduction by the oxidation of metals, increased osmotic pressure from high concentrations of mineral salts, and synergistic effects of metal ions also contribute to toxicity.[175] According to Parsons and Warner, in addition to the chemical effects of mine drainage, physical effects such as increased turbidity from soil erosion, accumulation of coal fines, and smothering of the stream substrate from precipitated metal compounds may also occur.[176]

---

[173]Kimmel, W. G., "The impact of acid mine drainage on the stream ecosystem." In *Pennsylvania Coal: Resources, Technology and Utilization,* Majumdar, S. K. and Miller, W. W. (eds.), *The Pa. Acad. Sci. Publ.,* pp. 424–437, 1983.
[174]Kimmel, W. G., "The impact of acid mine drainage on the stream ecosystem." In *Pennsylvania Coal: Resources, Technology and Utilization,* Majumdar, S. K. and Miller, W. W. (eds.), *The Pa. Acad. Sci. Publ.,* pp. 424–437, 1983.
[175]Parsons, J. D., "Literature pertaining to formation of acid mine waters and their effects on the chemistry and fauna of streams." *Trans. Ill. State Acad. Sci.,* v. 50, pp. 49–52, 1957.
[176]Parsons, J. D., "The effects of acid strip-mine effluents on the ecology of a stream." *Arch. Hydrobiol.,* 65:25–50, 1968; Warner, R.W., "Distribution of biota in a stream polluted by acid mine-drainage." *Ohio J. Sci.,* v. 71, pp. 202–215, 1971.

As mentioned, benthic macroinvertebrates are often used as indicators of water quality because of their varying degrees of sensitivity to pollutants. Unaffected streams generally have a variety of species with representatives of all insect orders, including a high diversity of insects such as mayflies, stoneflies, and caddisflies. Like many other potential pollutants, mine drainage can cause a reduction in the diversity and total numbers, or abundance, of macroinvertebrates and changes in community structure, such as a lower percentage of various macroinvertebrate taxa. Moderate pollution eliminates the more sensitive species.[177] Severely degraded conditions are characterized by dominance of certain taxonomic representatives of pollution-tolerant organisms, such as tubifex worms, midge larvae, alderfly larvae, fishfly larvae, cranefly larvae, caddisfly larvae, and non-benthic insects like predaceous diving beetles and water boatmen.[178] While these tolerant organisms may also be present in unpolluted streams, they dominate in impacted stream sections. Mayflies are generally sensitive to acid mine drainage; however, some stoneflies and caddisflies are tolerant of dilute acid mine drainage.

✓ *Note:* Most organisms have a well-defined range of pH tolerance. If the pH falls below the tolerance range, death will occur due to respiratory or osmoregulatory failure.

10.3.2 THERMAL POLLUTION

Oxygen is more soluble in cold water than in warm water; thus, oxygen levels are higher in colder waters. When a natural stream is heated by thermal pollution to a point above its normal water temperature, the stream's health is affected. The common source of thermal pollution is from cooling water of industrial power plants, which is discharged clean but quite warm into aquatic systems. Such thermal pollution has caused many complex aquatic problems. Mason reports that an "increase in temperature alters the physical environment, in terms of both a reduction in the density of the water and its oxygen concentration."[179] Moreover, Jeffries and Mills note that because all aquatic organisms have "thermal tolerance limits, a discharge may be lethal if beyond the threshold for a species."[180]

---

[177]Weed, C. E. and Rutschky, C. W., "Benthic macroinvertebrate community structure in a stream receiving acid mine drainage." *Proc. Pa. Acad. Sci.* v. 50, pp. 41–46, 1971.
[178]Nichols, L. E. and Bulow, F. J., "Effects of acid mine drainage on the stream ecosystem of the East Fork of the Obey River, Tennessee." *J. Tenn. Acad. Sci.,* v. 48, pp. 30–39, 1973; Roback, S.S. and Richardson, J. W., "The effects of acid mine drainage on aquatic insects." *Proc. Acad. Nat. Sci. Phil.,* v. 121, pp. 81–107, 1969; Parsons, J. D., "The effects of strip-mine effluents on the ecology of a stream." *Arch. Hydrobiol.,* v. 65, pp. 25–50, 1968.
[179]Mason, C. F., "Biological aspects of freshwater pollution." *Pollution: Causes, Effects, and Control,* Harrison, R.M. (ed.). Cambridge, Great Britain: The Royal Society of Chemistry, p. 118, 1990.
[180]Jeffries, M. and Mills, D., *Freshwater Ecology: Principles and Applications.* London: Belhaven Press, p. 178, 1990.

Thus, direct heat may cause the death of aquatic animals. Another effect is the increased susceptibility to toxins at higher solubility at higher temperatures. With the reduced solubility of oxygen in the water there is, in addition, an increase in the metabolism and respiratory demands of most animals because of higher temperatures, so that each animal actually requires more oxygen at 75°F than at 55°F.

Water temperature can be a principal ecological factor governing the presence, or absence, distribution, and abundance of aquatic life. A major increase in water temperature magnifies the effects of toxic and organic pollution, lowers the oxygen-holding capacity of water, and causes the death of many aquatic organisms.

✓ *Note:* The industrial use of large amounts of stream water for cooling is the primary cause of thermal pollution in streams. Thermal pollution is also caused by the removal of riparian vegetation. In fact, the most important factor influencing changes in stream water temperature is shade. In addition to shade provided by vegetation, water temperatures are also influenced by topography, surface area and volume of the stream, altitude, stream gradient, underground water inflow, and type of stream or channel. However, by maintaining adequate vegetation cover of such height and density as to adequately shade the stream during periods of maximum solar radiation, abnormal water temperature increases can often be prevented or minimized.[181]

10.3.3 CHEMICAL TOXINS

Several types of chemical toxins disrupt aquatic communities. The list includes phenol (toxic at 1.0 mg/L), arsenic (recommended limit of 0.01 mg/L), fluorides (recommended limit 0.9 mg/L), and cyanide (may be fatal to fish at 0.1 mg/L). Of particular concern to stream ecologists are those toxic compounds that accumulate in tissues, especially pesticides and PCBs. The problem these toxins pose to higher life forms, including humans, is that the toxins accumulate in tissues and are passed along the food chain.

10.3.4 HEAVY METALS

Heavy metals (metals generally in the first two columns of the periodic chart) are introduced into aquatic ecosystems as a result of the weathering of rocks and soils, volcanic activity, and a variety of anthropogenic (man-made) activities. Various heavy metals such as cadmium, copper, mercury, lead, silver, and chromium have also been found to be too toxic to aquatic life as well as

---

[181] Pope, P. E., *Forestry and Water Quality: Pollution Control Practices*. West Lafayette, IN: Purdue University, pp. 1–8, 2000.

human beings. Laws points out that "virtually all metals, including the essential metal micronutrients, are toxic to aquatic organisms as well as humans."[182]

It is interesting to note the findings of aquatic research relating to the impact on aquatic life of combining different metals in the same discharge. As a case in point, Smith points out that "many chemical wastes, harmless alone, react with other chemicals to produce highly toxic conditions."[183] Smith's point can be seen when small, harmless amounts of copper or zinc alone will not harm most aquatic organisms. However, when these two metals are combined (synergized), in even extremely small concentrations, they will destroy all the fish in a stream.

10.4 SELECTED INDICATORS OF STREAM WATER QUALITY[184]

Generally (from water year to water year), the most abundant data for describing U.S. stream water-quality conditions are traditional sanitary and chemical water-quality parameters (indicators) such as dissolved oxygen, fecal coliform bacteria, nutrients (nitrate and total phosphorus), dissolved solids, and suspended sediment (see Figure 10.2). [*Note:* A water year is the 12-month period from October 1 through September 30 and is identified by the calendar year in which it ends.]

10.4.1 DISSOLVED OXYGEN (DO)

Dissolved oxygen in streams is as critical to the good health of stream organisms as is gaseous oxygen to humans. DO is essential to the respiration of aquatic organisms, and its concentration in streams is a major determinant of the species composition of biota in the water and underlying sediments. Moreover, the DO in streams has a profound effect on the biochemical reactions that occur in water and sediments, which in turn affect numerous aspects of water quality, including the solubility of many lotic elements and aesthetic qualities of odor and taste. For these reasons, DO historically has been one of the most frequently measured indicators of water quality.[185]

In the absence of substances that cause its depletion, the DO concentration in stream water approximates the saturation level for oxygen in water in contact with the atmosphere and decreases with increasing water temperature from about 14 mg/L (milligrams per liter) at freezing to about 7 mg/L at 86°F (30°C).

---

[182]Laws, E. A., *Aquatic Pollution: An Introductory Text*. New York: John Wiley & Sons, Inc., p. 352, 1993.
[183]Smith, R. L., *Ecology and Field Biology*. New York: Harper & Row, p. 624, 1974.
[184]From USGS: Smith, R. A., Alexander, R. B., and Lanfear, K. J., *Stream Water Quality in the Conterminous United States-Status and Trends of Selected Indicators during the 1980's*. Washington, DC: U.S. Geological Survey (USGS) Water-Supply Paper 2400, pp. 1–12, February 1997.
[185]Hern, J. D., *Study and Interpretation of the Chemical Characteristics of Natural Water*, 3rd ed. Washington, DC: U.S. Geological Survey Water-Supply Paper 2254, p. 263, 1985.

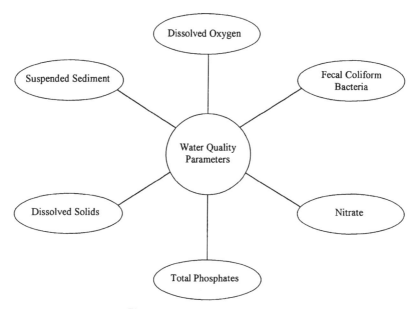

Figure 10.2 Water quality parameters.

For this reason, in ecologically healthy streams, the DO concentration depends primarily on temperature, which varies with season and climate.

Criteria for defining desirable DO concentration are often differentiated as applicable to cold-water biota, such as trout and their insect prey, and the more low-oxygen-tolerant species of warm-water ecosystems. Moreover, because of the critical respiratory function of DO in aquatic animals, criteria are often expressed in terms of the short-term duration and frequency of occurrence of minimum concentration rather than long-term average concentrations. Studies cited by the USEPA on the dependence of freshwater biota on DO suggest that streams in which the concentration is less than 6.5 mg/L for more than about 20% of the time generally are not capable of supporting trout or other cold-water fish, and such concentrations could impair population growth among some warm-water game fish, such as largemouth bass.[186]

Furthermore, streams in which the DO-deficit concentration is greater than 4 mg/L for more than 20% of the time generally cannot support either cold- or warm-water game fish. DO deficit refers to the difference between the saturation and measured concentrations of DO in a water sample and is a direct measure of the effects of oxygen-demanding substances on DO in streams.

Major sources of substances that cause depletion of DO in streams are dis-

---

[186]USEPA. *Quality Criteria for Water 1986.* Washington, DC: U.S. Environmental Protection Agency, Office of Water, EPA 440/5-86-00 [variously paginated], 1986.

charges from municipal and industrial wastewater treatment plants; leaks and overflows from sewage lines and septic tanks; stormwater runoff from agricultural and urban lands; and decaying vegetation, including aquatic plants from the stream itself and detrital terrestrial vegetation. DO is added to stream water by the process of aeration (waterfalls, riffles) and the photosynthesis of plants.

10.4.2 FECAL COLIFORM BACTERIA

Historically, the concentration of fecal coliform bacteria has been considered an important indicator of water quality because the presence of these organisms in streams is a reliable indicator of fecal contamination from warm-blooded animals.[187] The presence of fecal material in water where humans swim or where shellfish are harvested presents a significant risk of infection from pathogenic organisms associated with fecal coliform bacteria.[188] The correlation between documented cases of infectious disease and selected species of fecal coliform bacteria such as *Escherichia coli (E. coli)* is well established, but total concentrations of the fecal coliform group are easier to measure than selected species and, thus, have been widely used as indicators for many years. A concentration of 200 bacterial colonies/100 mL (colonies per 100 milliliters) of water has long been considered the acceptable limit for fecal-coliform density in waters where human contact occurs.[189] In addition to this limit, an arbitrary threshold of 1,000 bacterial colonies/100 mL was selected for this discussion to categorize high fecal-coliform concentrations. The major sources of fecal coliform bacteria are untreated sewage, effluent from sewage-treatment plants (point-source pollution), and runoff from pastures, feedlots, and urban areas (non-point-source pollution).

10.4.3 DISSOLVED SOLIDS

Dissolved solids refers to the sum of all dissolved constituents in a water sample. In most streams, the major components of dissolved solids are the ions of calcium, magnesium, sodium, potassium, sulfate, and chloride. The significance of these constituents in streams is related mostly to the potential limitations that large dissolved-solids concentrations impose on certain domestic, industrial, and irrigation water uses rather than to their ecological significance. In this discussion, dissolved-solids concentrations are classified arbitrarily as low (less than 100 mg/L), medium (100 to 500 mg/L), high (500 to 1000 mg/L), and very high (greater than 1000 mg/L).

---

[187]USEPA. *Quality Criteria for Water.* Washington, DC: U.S. Environmental Protection Agency, p. 256, 1976.
[188]USEPA. *Quality Criteria for Water.* Washington, DC: U.S. Environmental Protection Agency, EPA 440/5-86-00 [variously paginated], 1986.
[189]USEPA. *Quality Criteria for Water.* Washington, DC: U.S. Environmental Protection Agency, p. 256, 1976.

In most streams, the major source of dissolved solids is the dissolution of minerals naturally found in soil and rock. Because of the wide variation in the solubility of different minerals, and in the amount of precipitation available to dissolve them, the concentration of dissolved solids in streams nationwide ranges from only a few milligrams per liter to several thousand milligrams per liter. Within this broad range, the highest dissolved-solids concentrations (greater than 500 mg/L) are found in the arid Southwest, where high rates of evaporation and transpiration tend to concentrate dissolved solids. Concentrations are medium to high (greater than 100 mg/L) in parts of the midwestern United States, where soluble carbonates are abundant, and they are lowest (less than 100 mg/L) in the eastern and northwestern parts of the United States, where high precipitation rates dilute dissolved constituents. Average concentrations were lowest in forested areas and highest in range areas, which reflect the more arid areas of the country.

Human activities contribute significantly to dissolved-solids concentrations in most streams. For example, a moderate correlation between population and dissolved-solids concentration in streams has long been noted for much of the eastern and northwestern United States, where point-source municipal and industrial effluents typically have higher dissolved-solids concentrations than their receiving streams.[190] Also, land disturbance associated with mining and agriculture increases the exposure of mineral deposits to precipitation and increases the non-point-source load of dissolved solids. In recent years, the correlation between population and dissolved-solids concentrations has been strengthened by the increased dissolved salts in streams as a result of the increased use of highway deicing salt.[191]

10.4.4 NITRATE

Ecological concern about high concentrations of nitrate in streams stems from its potential for contributing to eutrophication, which is the excessive growth of aquatic plants that can impart unpleasant odors and tastes to water and reduce its clarity, and, upon dying, can lower the DO concentrations.[192] It has not been possible, however, to establish a nationally applicable threshold concentration for nitrate to protect against eutrophication because effects of nitrate concentrations are highly variable in different locations and are greatest in coastal waters that are far removed from inland nitrate sources. Historically,

---

[190]Peters, N. E., *Evaluation of Environmental Factors Affecting Yields of Major Dissolved Ions of Streams in the United States.* Washington, DC: U.S. Geological Survey Water-Supply Paper 2228, p. 39, 1984.
[191]Smith, R. A., Alexander, R. B., and Wolman, M. G., "Water-quality trends in the nation's rivers." *Science,* v. 235, no. 4796, pp. 1607–1615, 1987.
[192]National Academy of Sciences, *Eutrophication—Causes, Consequences, Correctives—Proceedings of International Symposium on Eutrophication,* Madison, Wisconsin, June 11–15, 1967. Washington, DC: National Academy of Sciences, p. 661, 1969.

government standards and eutrophication-control strategies for inland waters have focused on phosphorus concentration rather than on nitrate concentration because phosphorus is usually depleted more rapidly by the growth of aquatic plants than is nitrate, and, therefore, is frequently the limiting factor in eutrophication. Increasingly, however, it is recognized that control of estuarine and coastal eutrophication will require control of nitrate from inland sources.

Major sources of nitrate in streams are municipal and industrial wastewater discharge and agricultural and urban runoff. Deposition from the atmosphere of the nitrogenous material in automobile exhaust and industrial emissions is also a source.

10.4.5 TOTAL PHOSPHATE

In streams, phosphorus occurs primarily as phosphates and can be either dissolved, incorporated in organisms, or attached to particles in the water or in bottom sediments. Total phosphorus refers to the sum of all forms of phosphorus in a water sample and is reported in terms of elemental phosphorus.

Phosphorus is a particularly important nutrient in freshwater ecosystems because, as mentioned, it is usually the nutrient in shortest supply, and its availability often controls the rate of eutrophication. When human activities make phosphorus available in larger quantities, the accelerated growth of algae and other aquatic plants in streams can cause eutrophication, which depletes DO, imparts undesirable tastes and odors in water, and clogs water-supply intakes. To protect against eutrophication, the USEPA recommends an upper limit of 0.1 mg/L as the standard for total phosphorus in streams.[193] For this discussion, a threshold of 0.1 mg/L and an arbitrary threshold of 0.5 mg/L were selected for analysis of total phosphorus concentrations in streams.

Sources of phosphorus are the decomposition of organic matter and inorganic phosphate minerals that are mined and incorporated in fertilizers, detergents, and other commodities. Thus, major point sources of phosphorus to streams are waste discharges from wastewater treatment and food-processing plants and other industrial facilities. Non-point sources of phosphorus include agricultural and urban runoff and, in certain regions, the runoff and groundwater flow from areas that contain natural deposits of phosphate minerals.[194]

10.4.6 SUSPENDED SEDIMENT

The suspended-sediment concentration of streams consists of the total quantity of suspended organic and inorganic particulate matter in water and has an

---

[193]USEPA. *Quality Criteria for Water.* Washington, DC: U.S. Environmental Protection Agency, Office of Water, EPA 440/5-86-00 [variously paginated], 1986.
[194]Hem, J. D., *Study and Interpretation of the Chemical Characteristics of Natural Water,* 3rd ed. Washington, DC: U.S. Geological Survey Water-Supply Paper 2254, p. 263, 1985.

important influence on aspects of water use and ecosystem health. High concentrations of suspended sediment in streams diminish the recreational use of streams because pathogens and toxic substances commonly associated with suspended sediments are threats to public health. High concentrations also reduce water clarity, thereby affecting the aesthetic appeal of streams. They are detrimental to stream biota because they inhibit respiration and feeding, diminish the transmission of light needed for plant photosynthesis, promote infections, and, when the sediment is deposited, can suffocate benthic organisms, especially in the embryonic and larval stages.[195] Most sediment must be removed from water that is withdrawn for human use, and high sediment concentrations add significantly to the cost of treatment. Additionally, suspended sediment can cause significant wear to bridge footings and other stream structures, and, as it accumulates in a reservoir, it decreases its storage capacity.

The source of most suspended sediment is soil erosion. Although organic particles frequently form an important component of suspended sediment, most is inorganic by weight. Rates of soil erosion vary widely and depend on such factors as soil characteristics, precipitation frequency and intensity, slope of the land surface, and the nature and extent of land disturbance from agriculture, mining, and construction. Because the quantities of sediment entering streams depend greatly on natural factors, it is difficult to establish national criteria for suspended-sediment concentration. In many western areas, for example, stream ecosystems are naturally adapted to suspended-sediment concentrations that periodically are many times greater than those that are detrimental in other areas. Rather than establish national criteria for suspended-sediment concentration, the USEPA has recommended that light penetration in water not be reduced by suspended material by more than 10% from its natural level.[196] In this discussion, average suspended-sediment concentrations for this study period are arbitrarily grouped into three concentration classes—less than 100 mg/L, 100 to 500 mg/L, and greater than 500 mg/L.

10.5 SUMMARY OF KEY TERMS

- *Dissolved oxygen (DO)*—is found in amounts of 9 to 10 parts per million in streams and lakes.
- *Solubility of gaseous oxygen*—in water, controls the amount of DO in an aquatic system.
- *Oxygen*—is more soluble in cold water than in warm water (oxygen levels are higher in colder waters).
- *Oxygen demand*—from organisms and decaying organic matter increases as the temperature of the water increases.

---

[195]USEPA. *Quality Criteria for Water.* Washington, DC: U.S. Environmental Protection Agency, Office of Water, EPA 440/5-86-00 [variously paginated], 1986.
[196]Ibid, [variously paginated].

- *TMDL*—is essentially a "pollution budget" designed to restore the health of the polluted body of water.
- *pH*—refers to the concentration of hydrogen ions in water. The allowable pH range is 6.5 to 8.5 for protection of aquatic organisms and for controlling undesirable chemical reactions.
- *Thermal pollution*—refers to rising temperatures in streams that decrease DO content and increase the toxicity of substances by increasing their solubility or changing their ionic character.
- *Fecal coliform bacteria*—is commonly used as an indicator organism. That is, the presence of coliforms is taken as an indication that pathogenic organisms may be present, and the absence of coliforms is taken as an indication that the water is free from disease-producing organisms.
- *Dissolved solids*—consist of organic and inorganic molecules and ions that are present in true solution in water.

10.6 CHAPTER REVIEW QUESTIONS

10.1 The Clean Water Act mandates:

10.2 A TMDL is essentially a _____.

10.3 Examples of disease-causing agents include:

10.4 _____ are often used as indicators of water quality.

10.5 Oxygen is more soluble in _____ than in _____.

10.6 A major increase in water temperature magnifies the effect of _____ and _____ pollution.

10.7 The main problem _____ pose to human life forms is that the _____ accumulate in _____ and are passed along the _____.

10.8 List six indicators of water quality.

10.9 _____ historically has been one of the most frequently measured indicators of water quality.

10.10 _____ refers to the sum of all dissolved constituents in a water sample.

CHAPTER 11

Biomonitoring

In January, we take our nets to a no-name stream in the foothills of the Blue Ridge Mountains of Virginia to do a special kind of macroinvertebrate monitoring—looking for "winter stoneflies." Winter stoneflies have an unusual life cycle. Soon after hatching in early spring, the larvae bury themselves in the streambed. They spend the summer lying dormant in the mud, thereby avoiding problems like overheated streams, low oxygen concentrations, fluctuating flows, and heavy predation. In late November, they emerge, grow quickly for a couple of months, then lay their eggs in January.

January monitoring of winter stoneflies helps in interpreting the results of spring and fall macroinvertebrate surveys. In spring and fall, a thorough benthic survey is conducted, based on *Protocol II of the USEPA's Rapid Bioassessment Protocols for Use in Streams and Rivers*. Some sites on various rural streams have poor diversity and no sensitive families. Is the lack of macroinvertebrate diversity because of specific warm-weather conditions, high temperature, low oxygen, or fluctuating flows, or is some toxic contamination present? In the January screening, if winter stoneflies are plentiful, seasonal conditions were probably to blame for the earlier results; if winter stoneflies are absent, the site probably suffers from toxic contamination (based on our rural location, probably emanating from non-point sources) that is present year-round.

Though different genera of winter stoneflies are found in our region (southwestern Virginia), *Allocapnia* is sought because it is present even in the smallest streams.

11.1 WHAT IS BIOMONITORING?

THE life in, and physical characteristics of, a stream ecosystem provide insight into the historical and current status of its quality. The assessment of a stream ecosystem based on organisms living in it is called *biomonitoring*. The assessment of the system based on its physical characteristics is called a habitat assessment. Biomonitoring and habitat assessment are two tools that stream ecologists use to assess the water quality of a stream.

Biomonitoring is the process of inventorying aquatic organisms in a selected region of an aquatic system. The types of organisms can range from benthic macroinvertebrates to algae to fish. All organisms have a tolerance level for pollutants. Tolerance levels vary from organism to organism. Thus, it is possible to determine at a point in time, qualitatively, the level of water quality impairment based on what fauna are present and their abundances. It is also possible to monitor changes in water quality by repeated sampling of the same areas over an extended period of time. As mentioned, benthic macroinvertebrates are excellent indicators for several reasons:

(1) They are ubiquitous.
(2) They are relatively sedentary and long-lived (larval and nymph forms).
(3) Some species of benthic macroinvertebrates are sensitive to pollution, and some are tolerant.
(4) Benthic macroinvertebrates are easy to collect and identify.

Benthic macroinvertebrates act as continuous monitors of the water they live in. Unlike chemical monitoring, which provides information about water quality at the time of measurement (a snapshot), biological monitoring can provide information about past and/or episodic pollution (a videotape). This concept is analogous to miners who took canaries into deep mines with them to test for air quality. If the canary died, the miners knew the air was bad, and they had to leave the mine. Biomonitoring a stream ecosystem uses the same theoretical approach. Aquatic macroinvertebrates are subject to pollutants in the stream. Consequently, the health of the organisms reflects the quality of the water they live in. If the pollution levels reach a critical concentration, certain organisms will migrate away, fail to reproduce, or die, eventually leading to the disappearance of those species at the polluted site. Normally, these organisms will return if conditions improve in the system.[197]

Biomonitoring (and the related term, bioassessment) surveys are conducted before and after an anticipated impact to determine the effect of the activity on the stream habitat. Moreover, surveys are performed periodically to monitor stream habitats and watch for unanticipated impacts. Finally, biomonitoring surveys are designed to reference conditions or to set biocriteria (serve as monitoring thresholds to signal future impacts, regulatory actions, etc.) for determining that an impact has occurred.[198]

✓ *Note:* The primary justification for bioassessment and monitoring is that degradation of stream habitats affects the biota using those habitats, and, therefore, the living organisms themselves provide the most direct means of assessing real environmental impacts.

[197]Byl, T. D. and Smith, G. F., *Biomonitoring Our Streams: What's It All About?* Nashville, TN: U.S. Geological Survey, pp. 2–3, 1994.
[198]Camann, M., *Freshwater Aquatic Invertebrates: Biomonitoring.* www.humboldt.edu, pp. 1–4, 1996.

11.2 BIOTIC INDEX

Certain common aquatic organisms, by indicating the extent of oxygenation of a stream, may be regarded as indicators of the intensity of pollution from organic waste. The responses of aquatic organisms in streams to large quantities of organic wastes are well documented. They occur in a predictable cyclical manner. For example, upstream from the discharge point, a stream can support a wide variety of algae, fish, and other organisms, but in the section of the stream where oxygen levels are low (below 5 ppm), only a few types of worms survive. As stream flow courses downstream, oxygen levels recover, and those species that can tolerate low rates of oxygen (such as gar, catfish, and carp) begin to appear. Eventually, at some further point downstream, a clean water zone reestablishes itself, and a more diverse and desirable community of organisms returns.

During this characteristic pattern of alternating levels of dissolved oxygen (in response to the dumping of large amounts of biodegradable organic material), a stream, as stated above, goes through a cycle called an *oxygen sag curve*. Its state can be determined using the biotic index as an indicator of oxygen content. The oxygen sag curve will be discussed in greater detail later.

Biological monitoring has evolved rapidly during the twentieth century as knowledge of stream ecosystems has changed, and human-imposed stresses have become more complex and pervasive. Early water quality specialists developed biotic indices sensitive to organic pollution (sewage) and sedimentation; this approach continues in modern "biotic indices."[199]

The biotic index is a systematic survey of macroinvertebrate organisms. Because the diversity of species in a stream is often a good indicator of the presence of pollution, the biotic index can be used to correlate with stream quality. Observation of types of species present or missing is used as an indicator of stream pollution. The biotic index, used in the determination of the types, species, and numbers of biological organisms present in a stream, is commonly used as an auxiliary to BOD determination in determining stream pollution.

The biotic index is based on two principles:

(1) A large dumping of organic waste into a stream tends to restrict the variety of organisms at a certain point in the stream.

---

[199]Kolkwitz, R. and Marsson, M., *Okologie der pflanzlichen saprobiean. Berichte der Deutchen botanischen Gesellschaft,* 26a:505–519, 1906. (Translated 1967. "Ecology of plant saprobia," pp. 47–52 in Kemp, L. E., Ingram, W. M., and Mackenthum, K. M. (eds). *Biology of Water Pollution.* Washington, DC: Federal Water Pollution Control Administration; Clutter, F. M., "An empirical biotic index of the quality of water in South African streams and rivers." *Water Research,* 6:19–30, 1972; Hilsenhoff, W. L., "Using a biotic index to evaluate water quality in streams." *Technical bulletin number 132.* Madison, WI, p. 21, 1982; Lenat, D., "Water quality assessment of streams using a qualitative collection method for benthic macroinvertebrates." *Journal North American Benthological Society,* 7:222–233, 1988; Lenat, D., "A biotic index for the southeastern United States: derivation and list of tolerance values, with criteria for assigning water quality ratings." *Journal North American Benthological Society,* 12:279–290, 1993.

(2) As the degree of pollution in a stream increases, key organisms tend to disappear in a predictable order. The disappearance of particular organisms tends to indicate the water quality of the stream.

There are several different forms of the biotic index. In Great Britain, for example, the Trent Biotic Index (TBI), the Chandler score, the Biological Monitoring Working Party (BMWP) score, and the Lincoln Quality Index (LQI) are widely used. Most of the forms use a biotic index that ranges from 0 to 10. The most polluted stream, which therefore contains the smallest variety of organisms, is at the lowest end of the scale (0); the clean streams are at the highest end (10). A stream with a biotic index of greater than 5 will support game fish; on the other hand, a stream with a biotic index of less than 4 will not support game fish.

Because they are easy to sample, macroinvertebrates have predominated in biological monitoring. In addition, macroinvertebrates can be easily identified using identification keys that are portable and easily used in field settings. Present knowledge of macroinvertebrate tolerances and responses to stream pollution is well documented. In the United States, for example, the Environmental Protection Agency (EPA) has required states to incorporate a narrative biological criteria into its water quality standards by 1993. The National Park Service (NPS) has collected macroinvertebrate samples from American streams since 1984. Through their sampling effort, NPS has been able to derive quantitative biological standards.[200]

Macroinvertebrates are a diverse group. They demonstrate tolerances that vary between species. Thus, discrete differences tend to show up, containing both tolerant and sensitive indicators.

The biotic index provides a valuable measure of pollution. This is especially the case for species that are very sensitive to lack of oxygen. An example of an organism that is commonly used in biological monitoring is the stonefly. Stonefly larvae live underwater and survive best in well-aerated, unpolluted waters with clean gravel bottoms. When stream water quality deteriorates due to organic pollution, stonefly larvae cannot survive. The degradation of stonefly larvae has an exponential effect upon other insects and fish that feed off the larvae; when the stonefly larvae disappears, so in turn do many insects and fish.[201]

Table 11.1 shows a modified version of the BMWP biotic index. Considering that the BMWP biotic index indicates ideal stream conditions, it takes into account the sensitivities of different macroinvertebrate species to stream contamination. As mentioned, aquatic macroinvertebrate species are repre-

---

[200]Huff, W. R., "Biological indices define water quality standard." *Water Environment & Technology,* 5, pp. 21–22, 1993.
[201]O'Toole, C. (ed.), *The Encyclopedia of Insects.* New York, Facts on File, Inc., p. 134, 1986.

TABLE 11.1. The BMWP Score System (Modified for Illustrative Purposes).

Families	Common-Name Examples	Score
Heptageniidae	Mayflies	10
Leuctridae	Stoneflies	
Aeshnidae	Dragonflies	8
Polycentropidae	Caddisflies	7
Hydrometridae	Water strider	
Gyrinidae	Whirligig beetle	5
Chironomidae	Mosquitoes	2
Oligochaera	Worms	1

sented by diverse populations and are excellent indicators of pollution. These aquatic macroinvertebrates are organisms that are large enough to be seen by the unaided eye. Moreover, most aquatic macroinvertebrate species live for at least a year; and they are sensitive to stream water quality both on a short-term basis and on a long-term basis. For example, mayflies, stoneflies, and caddisflies are aquatic macroinvertebrates that are considered to be clean-water organisms; they are generally the first to disappear from a stream if water quality declines and are, therefore, given a high score. On the other hand, tubificid worms (which are tolerant to pollution) are given a low score.

In Table 11.1, a score from 1–10 is given for each family present. A site score is calculated by adding the individual family scores. The site score or total score is then divided by the number of families recorded to derive the Average Score Per Taxon (ASPT). High ASPT scores result due to such taxa as stoneflies, mayflies, and caddisflies being present in the stream. A low ASPT score is obtained from streams that are heavily polluted and dominated by tubificid worms and other pollution-tolerant organisms.

From Table 11.1, it can be seen that those organisms having high scores, especially mayflies and stoneflies, are the most sensitive, and others, such as dragonflies and caddisflies, are very sensitive to any pollution (deoxygenation) of their aquatic environment.

11.3 BENTHIC MACROINVERTEBRATE BIOTIC INDEX

The Benthic Macroinvertebrate Biotic Index employs the use of certain benthic macroinvertebrates to determine (or gauge) the water quality (relative health) of a stream.

In this discussion, benthic macroinvertebrates are classified into three groups based on their sensitivity to pollution. The number of taxa in each of these groups will be tallied and assigned a score. The scores are then summed to yield a score that can be used as an estimate of the quality of the stream for life.

TABLE 11.2. Sample Index of Macroinvertebrates.

Group One (Sensitive)	Group Two (Somewhat Sensitive)	Group Three (Tolerant)
Stonefly larva	Alderfly larva	Aquatic worm
Caddisfly larva	Damselfly larva	Midgefly larva
Water penny larva	Cranefly larva	Blackfly larva
Riffle beetle adult	Beetle adult	Leech
Mayfly larva	Dragonfly larva	Snails

11.3.1 METRICS WITHIN THE BENTHIC MACROINVERTEBRATES

The three groups based on the sensitivity to pollution are described as follows:

(1) *Group One—Pollution-Sensitive Organisms:* These organisms are highly sensitive to pollution.
(2) *Group Two—Pollution-Intermediate Organisms:* These organisms are somewhat less sensitive to pollution, but they cannot live in very polluted water.
(3) *Group Three—Pollution-Intolerant Organisms:* These organisms are rather insensitive to pollution. While they require basic resources for life, they can be found in fairly polluted water.

A sample index of macroinvertebrates, in regard to sensitivity to pollution, is listed in Table 11.2.

In summarizing the use of biological data as indicators of polluted or unpolluted streams, it can be said that unpolluted streams normally support a wide variety of macroinvertebrates and other aquatic organisms with relatively few of any one kind. Any significant change in the normal population usually indicates pollution.

11.4 SUMMARY OF KEY TERMS

- *Biomonitoring*—refers to the systematic use of living organisms or their responses to determine the quality of the aquatic environment.
- *Biotic indices*—classify the degree of pollution by determining the tolerance or sensitivity of taxonomically homogeneous organisms or several groups of indicator organisms to a given pollutant.

11.5 CHAPTER REVIEW QUESTIONS

11.1 Any significant change in the normal population of a stream usually indicates _____.

11.2 _____ is the process of inventorying aquatic organisms in a selected region of an aquatic system.

11.3 List four reasons why benthic macroinvertebrates are excellent indicators.

11.4 The _____ is a systematic survey of macroinvertebrate organisms.

11.5 The biotic index is based on what two principles?

11.6 In biomonitoring a stream where caddisfly larva and mayfly and stonefly nymphs are present, the stream's water quality is most likely _____.

11.7 In biomonitoring a stream where aquatic worms and snails predominate, the water quality is most likely _____.

11.8 Stonefly and mayfly nymphs are _____ to pollution.

11.9 Beetle and cranefly larva are somewhat _____ tolerant.

11.10 Snails, leeches, and midge larva are _____ tolerant.

CHAPTER 12

Self-Purification of Streams

In terms of practical usefulness the waste assimilation capacity of streams as a water resource has its basis in the complex phenomenon termed stream self-purification. This is a dynamic phenomenon reflecting hydrologic and biologic variations, and the interrelations are not yet fully understood in precise terms. However, this does not preclude applying what is known. Sufficient knowledge is available to permit quantitative definition of resultant stream conditions under expected ranges of variation to serve as practical guides in decisions dealing with water resource use, development, and management.—C. J. Velz[202]

12.1 BALANCING THE "AQUARIUM"

AN outdoor excursion to the local stream can be a relaxing and enjoyable undertaking. On the other hand, when you arrive at the local stream and look upon the stream's flowing mass to discover a parade of waste and discarded rubble bobbing along the stream's course and cluttering the adjacent shoreline and downstream areas, any feeling of relaxation or enjoyment is quickly extinguished. Further, the sickening sensation the observer feels is made worse as closer scrutiny of the putrid flow is gained. The rainbow-colored shimmer of an oil slick, interrupted here and there by dead fish and floating refuse, and the slimy fungal growth that prevails are recognized. At the same time, the observer's sense of smell is alerted to the noxious conditions. Along with the fouled water and the stench, the observer notices signs warning, "DANGER—NO SWIMMING or FISHING." The observer has discovered what ecologists have known and warned about for years. That is, contrary to popular belief, rivers and streams do not have an infinite capacity for pollution.

Before the early 1970s, such disgusting occurrences as the one just described were common along the rivers and streams near main metropolitan ar-

---

[202]Velz, C. J., *Applied Stream Sanitation*. New York: Wiley-Interscience, p. 66, 1970.

eas throughout most of the United States. Many aquatic habitats were fouled during the past because of industrialization. However, our streams and rivers were not always in such deplorable condition.

Before the Industrial Revolution of the 1800s, metropolitan areas were small and sparsely populated. Thus, river and stream systems within or next to early communities received insignificant quantities of discarded waste. Early on, these river and stream systems were able to compensate for the small amount of wastes they received. They have the ability to restore themselves through their own self-purification process. It was only when humans gathered in great numbers to form cities that the stream systems were not always able to recover from having received great quantities of refuse and other wastes.

Halsam pointed out that man's actions are determined by his expediency. We have the same amount of water as we did millions of years ago, and through the water cycle, we continually reuse that same water—water that was used by the ancient Romans and Greeks is the same water being used today. Increased demand by man has put enormous stress on our water supply. Thus, man upsets the delicate balance between pollution and the purification process of rivers and streams, unbalancing the "aquarium."

With the advent of industrialization, local rivers and streams became deplorable cesspools that worsened with time. During the Industrial Revolution, the removal of horse manure and garbage from city streets became a pressing concern; for example, Moran et al. point out that "none too frequently, garbage collectors cleaned the streets and dumped the refuse into the nearest river."[203] Halsam reports that as late as 1887, river keepers gained full employment by removing a constant flow of dead animals from a river in London. Moreover, the prevailing attitude of that day was "I don't want it anymore, throw it into the river."[204]

As of the early 1970s, any threat to the quality of water destined for use for drinking and recreation has quickly angered those affected. Fortunately, since the 1970s, efforts have been made to correct the stream pollution problem. Through scientific study and incorporation of wastewater treatment technology, streams have begun to be restored to their natural condition. And, the stream itself aids in restoring its natural water quality through the phenomenon of self-purification.

A balance of biological organisms is normal for all streams. Clean, healthy streams have certain characteristics in common. For example, one property of streams is their ability to dispose of small amounts of pollution. However, if streams receive unusually large amounts of waste, the stream life will change and attempt to stabilize such pollutants; that is, the biota will attempt to balance

---

[203]Moran, J. M., Morgan, M. D., and Wiersma, J. H., *Introduction to Environmental Science*. New York: W.H. Freeman and Company, p. 211, 1986.
[204]Halsam, S. M., *River Pollution: An Ecological Perspective*. New York: Belhaven Press, p. 21, 1990.

the "aquarium." However, if the stream biota are not capable of self-purifying, then the stream may become a lifeless body.

The self-purification process discussed here relates to the purification of organic matter only. In this chapter, organic stream pollution and the self-purification process will be discussed.

12.2 SOURCES OF STREAM POLLUTION

Sources of stream pollution are normally classified as point or non-point sources. A *point source* (PS) is a source that discharges effluent, such as wastewater from sewage treatment and industrial plants. A point source is usually easily identified as "end of the pipe" pollution; that is, it emanates from a concentrated source or sources. In addition to organic pollution received from the effluents of sewage treatment plants, other sources of organic pollution include runoffs and dissolution of minerals throughout an area and are not from one or more concentrated sources.

Non-concentrated sources are known as non-point sources (see Figure 12.1). *Non-point source* (NPS) pollution, unlike pollution from industrial and sewage treatment plants, comes from many diffuse sources. NPS pollution is caused by rainfall or snowmelt moving over and through the ground. As the

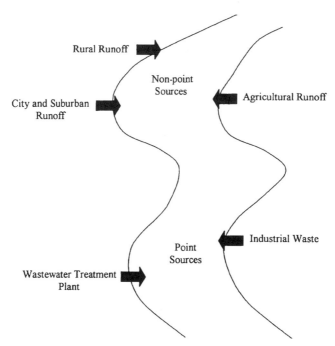

Figure 12.1 Point and non-point sources of pollution.

runoff moves, it picks up and carries away natural and man-made pollutants, finally depositing them into streams, lakes, wetlands, rivers, coastal waters, and even our underground sources of drinking water. These pollutants include the following:

- excess fertilizers, herbicides, and insecticides from agricultural lands and residential areas
- oil, grease, and toxic chemicals from urban runoff and energy production
- sediment from improperly managed construction sites, crop and forest lands, and eroding streambanks
- salt from irrigation practices and acid drainage from abandoned mines
- bacteria and nutrients from livestock, pet wastes, and faulty septic systems

Atmospheric deposition and hydromodification are also sources of non-point source pollution.[205]

As mentioned, specific examples of non-point sources include runoff from agricultural fields and also cleared forest areas, construction sites, and roadways. Of particular interest to environmentalists in recent years has been agricultural effluents. As a case in point, farm silage effluent has been estimated to be more than 200 times as potent [in terms of biochemical oxygen demand (BOD)] as treated sewage.[206]

Nutrients are organic and inorganic substances that provide food for microorganisms such as bacteria, fungi, and algae. Nutrients are supplemented by the discharge of sewage. The bacteria, fungi, and algae are consumed by the higher trophic levels in the community. Each stream, due to a limited amount of dissolved oxygen (DO), has a limited capacity for aerobic decomposition of organic matter without becoming anaerobic. If the organic load received is above that capacity, the stream becomes unfit for normal aquatic life, and it is not able to support organisms sensitive to oxygen depletion.[207]

Effluent from a sewage treatment plant is most commonly disposed of in a nearby waterway. At the point of entry of the discharge, there is a sharp decline in the concentration of DO in the stream. This phenomenon is known as the *oxygen sag*. Unfortunately (for the organisms that normally occupy a clean, healthy stream), when the DO is decreased, there is a concurrent massive increase in BOD as microorganisms utilize the DO as they break down the organic matter. When the organic matter is depleted, the microbial population and BOD decline, while the DO concentration increases, assisted by stream

---

[205]USEPA. *What is Nonpoint Source Pollution?* Washington, DC: United States Environmental Protection Agency, EPA-F-94-005, pp. 1–5, 1994.
[206]Mason, C. F., "Biological aspects of freshwater pollution." In *Pollution: Causes, Effects, and Control*. Harrison, R.M. (ed.), Cambridge, Great Britain: The Royal Society of Chemistry, p. 113, 1990.
[207]Smith, R. L., *Ecology and Field Biology*. New York: Harper & Row, p. 323, 1974.

flow (in the form of turbulence) and by the photosynthesis of aquatic plants. This self-purification process is very efficient, and the stream will suffer no permanent damage as long as the quantity of waste is not too high. Obviously, an understanding of this self-purification process is important to prevent overloading of the stream ecosystem.

As urban and industrial centers continue to grow, waste disposal problems also grow. Because wastes have increased in volume and are much more concentrated than before, natural waterways must have help in the purification process. This help is provided by wastewater treatment plants. A wastewater treatment plant functions to reduce the organic loading that raw sewage would impose on discharge into streams. Wastewater treatment plants utilize three stages of treatment: primary, secondary, and tertiary treatment. In breaking down the wastes, a secondary wastewater treatment plant uses the same type of self-purification process found in any stream ecosystem. Small bacteria and protozoans (one-celled organisms) begin breaking down the organic material. Aquatic insects and rotifers are then able to continue the purification process. Eventually, the stream will recover and show little or no effects of the sewage discharge. This phenomenon is known as *natural stream purification*.[208]

12.3 SAPROBITY OF A STREAM

Treated or untreated sewage dumped into streams can upset the ecological stability of the stream. Through natural processes and bacterial activity, streams can purify themselves. High concentrations of organic substances encourage the growth of decomposers such as bacteria and fungi, which convert the biodegradable organic substances in the stream into their cells and into basic substances like carbon dioxide, nitrates, sulfates, and phosphates. These basic substances and those contributed by the dissolution of rocks are converted by producers, algae and other plants, into their protoplasm. The normal food chain is then established with higher trophic levels. All consumers produce wastes that, with the organics from runoffs and sewage, are converted by bacteria and fungi into basic substances, thus establishing an ecosystem or a cyclic phenomenon.

Excess organic wastes upset this system by depleting the dissolved oxygen (DO) required by bacteria for aerobic decomposition of organics. In other words, the biochemical oxygen demand (BOD) of the stream increases, creating an inverse relationship between sewage and oxygen in the stream. The normal amount of dissolved oxygen in streams is above 9 mg/L at 20°C (68°F) water temperature. As the level of DO decreases to 5 mg/L, sensitive organisms—such as predators like trout—disappear. Figure 12.2 shows the

---

[208]Spellman, F. R. and Whiting, N. E., *Water Pollution Control Technology: Concepts and Applications.* Rockville, MD: Government Institutes, pp. 247–317, 1999.

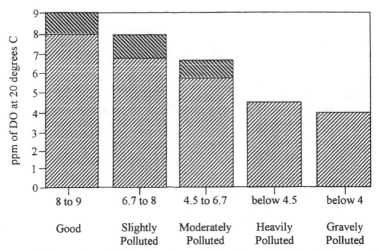

Figure 12.2 Water quality and DO content. (Source: Adapted from G. T. Miller, *Environmental Science: An Introduction.* Belmont, CA: Wadsworth, p. 351, 1988.)

correlation between water quality and dissolved oxygen (DO), in parts per million at 20°C.

As oxygen depletion progresses, other game fish, insects, crustaceans, rotifers, and even sensitive protozoans tend to be absent from the food chains. Ultimately, bacteria of facultative (can use oxygen and, under certain conditions, can grow in the absence of oxygen) and anaerobic types exist. Due to reaeration, streams do not reach a 0 ppm DO level and, thus, seldom go anaerobic. The degree of pollution and the character of the stream determine the amount of time the self-purification process will take.

The amount of organic matter and the activity by microbial communities living on it is called the *saprobity* of the stream's ecosystem. The term saprobity was introduced in Germany early in the twentieth century for the assessment of water quality, and saprobity as both a term and practical approach has been primarily used in Europe. Waters are said to have saprobic level (which can be measured using the species present and their relative abundance), in effect, a biotic index of organic pollution. As mentioned, the communities change, qualitatively and quantitatively, as organic content increases.[209]

12.3.1 DEFINITION OF KEY TERMS

In order to better appreciate a discussion of stream saprobity (i.e., stream

---

[209] Adapted from Jeffries, M. and Mills, D., *Freshwater Ecology: Principles and Applications.* London: Belhaven Press, p. 154, 1990.

pollution) and the self-purification process, a restatement, in greater detail, of two critical terms, previously defined or mentioned, is necessary:

(1) *Dissolved oxygen (DO)* is the amount of oxygen dissolved in a stream. It indicates the degree of health of the stream and its ability to support a balanced aquatic ecosystem. The oxygen comes from the atmosphere by solution and from photosynthesis of water plants. In a lentic (lake) environment, oxygen is added primarily by photosynthetic activity and secondarily by wind-induced wave action. In fast streams, oxygen is added primarily through reaeration from the atmosphere in rapids, waterfalls, and cascades. DO concentrations are usually higher and more uniform from surface to bottom in streams than in lakes.

(2) *Biochemical oxygen demand (BOD)* is the amount of oxygen required to biologically oxidize organic waste matter over a stated period of time. BOD is important in the self-purification process, because in order to estimate the rate of deoxygenation in the stream, the five-day and ultimate BOD must be known.

Most sewage wastes contain high concentrations of organic substances. Their presence encourages the growth of decomposers. Decomposers consume large quantities of DO.

A stream receiving an excessive amount of sewage (organic wastes) exhibits changes, which can be differentiated and classified into zones. Upstream, before a single point of pollution discharge, the stream is defined as having a *clean zone*. At the point of discharge, the water becomes turbid. This is called the *zone of recent pollution*. Shortly below the discharge point, the level of dissolved oxygen falls sharply and, in some cases, may fall to zero; this is called the *septic zone* (Figure 12.3).

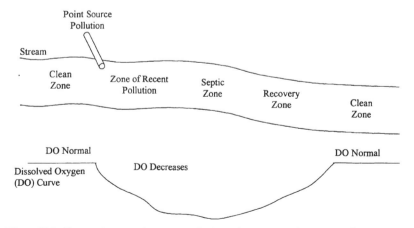

Figure 12.3 Changes that occur in a stream after it receives an excessive amount of raw sewage.

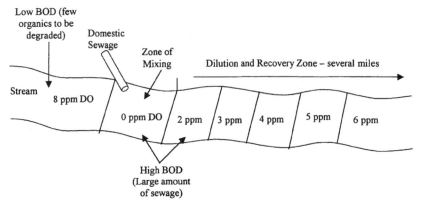

Figure 12.4 Effect of organic wastes on DO. (Source: Adapted from E. Enger, J. R. Kormelink, B. F. Smith, and R. J. Smith, *Environmental Science: The Study of Interrelationships*. Dubuque, IA: William C. Brown Publishers, p. 411, 1989.)

After the organic waste has been largely decomposed, the dissolved oxygen level begins to rise in the *recovery zone*. Eventually, given enough time and no further waste discharges, the stream will return to conditions similar to those in the clean zone.

The total change in organic matter in the stream at any time can be modeled. One simple model makes the assumption that the total change in the concentration of organic matter per time is a function of the initial rate of input of organic matter minus losses due to in-stream decomposition, assimilation by detritivores, and sedimentation of waste.[210]

In Figure 12.4 it can clearly be seen that sewage containing a high concentration of organic material is attacked by organisms, which use oxygen in the degradation process. Thus, there is an inverse relationship between oxygen and sewage in the stream. The greater the BOD, the less desirable the stream is for human use.

As stated previously, when excessive sewage is dumped into a stream, change occurs. These changes are shown in Figure 12.2. In order to foster a better appreciation for the changes that occur in each zone, the following information is provided.

12.3.1.1 Clean Water Zone

The clean water zone (see Figure 12.3) is the stretch of stream above the point of discharge (and is restored downstream once the self-purification pro-

---

[210] Westman, W. E., *Ecology, Impact Assessment, and Environmental Planning*. New York: John Wiley & Sons, Inc., p. 233, 1985.

cess is complete). In this zone, the stream is in an entirely natural state and contains no pollutants. Many different organisms are present, including the mayfly nymph, which has a narrow range of tolerance for DO. Also, many kinds of game fish are present in this zone. The following is a list of other characteristics:

(1) High DO
(2) Low BOD
(3) Clear water (low turbidity) and no odors
(4) Low bacterial count
(5) Low organic content
(6) High species diversity
(7) Bottom clean and free of sludge
(8) Presence of normal communities containing sensitive organisms such as bass, bluegill, perch, crayfish, and stonefly nymphs

12.3.1.2 Zone of Recent Pollution (Degradation Zone)

The zone of recent pollution (see Figure 12.3) occurs at the point of sewage discharge where turbidity increases while the DO content decreases. This sudden introduction of a heavy load of sewage (organic pollution) increases BOD and, hence, accelerates the growth of bacteria and fungi. When the organic material is degraded by organisms, the amount of DO decreases in various points in a stream and leads to a succession of changes in community structure. Changes caused by the pollution in the environment and the community are as follows:

(1) DO variable depending upon organic load
(2) High BOD
(3) Turbidity high
(4) Bacterial count high and increasing
(5) Lower species diversity
(6) Increase in number of individuals per species
(7) Appearance of slime molds and sludge deposits on bottom

The biota is represented by the following:

(1) *Flora* (Plants): blue-green algae, spirogyra, gomphonema
(2) *Annelids:* sludgeworms (Tubificidae)
(3) *Insects:* back swimmers, water boatman, and dragonflies
(4) *Fish:* tolerant fish such as catfish, gars, and carp

12.3.1.3 Septic Zone (Active Decomposition)

At this stage, active decomposition of the organic matter is proceeding at the optimum rate; thus, the rate of deoxygenation is greater than the supply or reaeration rate from the atmosphere. In some cases, DO is completely absent, hence the name *septic zone* (see Figure 12.3). In this zone, the organic waste material requires more oxygen in its decomposition than is naturally available in the stream. Only a few species other than bacteria occupy this zone. For example, in general, fish are completely absent. If the organic load is too high, bacteria may consume all the DO and start anaerobic decomposition of organics by first obtaining oxygen from nitrates and sulfates and then continuing without any oxygen. Anaerobic products include hydrogen sulfide, ammonia, methane, and hydrogen, which cause offensive odors (H_2S causes rotten-egg odor) and a toxic environment. Sludge mats may form and rising gas bubbles result. Due to reaeration, streams normally do not go completely septic (anaerobic). The rate of reaeration increases with the decrease in dissolved oxygen in the water and vice versa. Other characteristics may include the following:

(1) Very little to the complete absence of DO, especially during warm weather
(2) BOD high but decreasing
(3) Water very turbid and dark, often with an offensive odor
(4) High but decreasing organic content
(5) High bacterial count
(6) Low species diversity
(7) Slime blanket on the bottom with floating sludge
(8) Oily appearance on the water surface
(9) Rising gas bubbles

The biota present is represented by the species that are highly adapted to polluted conditions:

(1) *Flora:* only some blue-green algae
(2) *Annelids:* sludgeworms
(3) *Insects:* mosquito larvae and rattailed larvae (drone flies)
(4) *Mollusks:* air-breathing snails
(5) *Fish:* absent

12.3.1.4 Recovery Zone

In the recovery zone (see Figure 12.3), the stream has nearly completed its self-purification process. Most of the organic matter has been decomposed into basic substances such as nitrates, sulfates, and carbon dioxide. A gradual recovery of the stream occurs, due to reaeration, as the water gradually clears. It has a green-like tinge due to the growth of algal planktons. The algal growth is encouraged by increased transparency and availability of nitrates and sulfates. Many aquatic organisms that have a narrow range of tolerance for dissolved oxygen begin to appear in the stream. Conditions and biota of the recovery zone can be summarized as follows:

(1) DO content may range from 2 ppm to saturation value depending on the recovery stage
(2) Lower BOD
(3) Water less turbid and lighter in color, with decreasing odor
(4) Number of bacteria decreasing
(5) Lower organic content
(6) Number of species increasing and number of each species decreasing
(7) Less slime on the bottom with some sludge deposits
(8) Biota is characterized at first by the tolerant species, like those present in recent pollution zone; then by the appearance of some of the clean-water types
   - *Flora:* blue-green algae, phytoflagellates such as euglena, chlorophytes cholorella, and spirogyra
   - *Insects:* blackfly larvae and giant water bugs
   - *Mollusks:* clams
   - *Fish:* catfish and sunfish

The extent of complete recovery of a stream varies depending on "the stream's volume, flow rate, and the volume of incoming biodegradable wastes."[211] There used to be a common saying that every stream recovers within 30 miles from the point of organic pollution. This is not true. In this modern age, the actions of human beings have changed the character of most streams. Through diversion of stream channels and construction of dams, streams have lost some or most of their dilution ability. Moreover, the addition of more exotic types of nonbiodegradeable materials in the form of industrial wastes has affected a stream's ability to self-purify itself. As Enger et al. point out, because of the increasing amounts of industrial wastes that have been produced and dumped into our streams, the federal government has enacted legislation, such as various amendments to the Federal Water Pollution Control Act [Clean Water Act], mandating changes in how industry treats water. Basically, the Federal Act requires industries to treat industrial wastewater prior to it being returned to its source.[212]

---

[211]Miller, G. T., *Environmental Science: An Introduction.* Belmont, CA: Wadsworth, p. 351, 1988.
[212]Enger, E., Kormelink, J. R., Smith, B. F., and Smith, R. J., *Environmental Science: The Study of Interrelationships.* Dubuque, IA: William C. Brown Publishers, p. 312, 1981.

12.4 ORGANISMS AND THEIR ROLE IN SELF-PURIFICATION

As mentioned, the self-purification process in streams is similar to the purification process of secondary sewage treatment; that is, biological and chemical processes are used to remove most of the organic matter.[213] Secondary wastewater treatment is analogous to a "stream in a box."

✓ *Note:* In this discussion of self-purification of streams, it is the biological process that is being addressed.

When discussing the biological self-purification of streams, it is prudent to begin with the indicators of water quality. Four indicators of water quality are the coliform bacteria count, concentration of DO, BOD, and the Biotic Index. The biota that exist at various stages in the self-purification of a stream are direct indicators (a biotic index) of the condition of the water. Based on our experience, this biotic index is often more reliable than the chemical tests.

Aquatic organisms are responsible for degrading or decomposing organic wastes. Both the sewage treatment plant and the stream exhibit a change in the type of organisms present as the strength of the waste decreases. As the organic wastes are received by the stream, a large number of bacteria predominate because they thrive on the energy they receive from the organic waste. Some of these bacteria are normally found in streams. Others, such as enteric bacteria (coliform bacteria, found in great numbers in the intestines and thus in the feces of humans and other animals), are in a strange environment. The growth of normal stream bacteria is greatly enhanced by the organic nutrients. However, coliforms and pathogens generally die out within a few days, perhaps due to predation and unfavorable conditions. The bacteria predominate during the recent pollution zone and to near the end of the septic zone. If the organic load is too high, then the bacterial type changes from aerobic (those requiring oxygen) to anaerobic (those not requiring oxygen), due to the similar changes in conditions.

As stabilization continues, bacterial food becomes limited due to its high populations, and protozoans increase and eventually predominate. The protozoans are one-celled and feed on bacteria. Examples of protozoa are amoeba, paramecium, and other ciliates. As the food supply diminishes, protozoans decrease in population and are consumed by rotifers (wheel animalcules) (see Figure 12.5) and crustaceans in the recovery zone. During this period, turbidity has deceased and algal growth is stimulated.

There is also a change in the type of aquatic insects present in a polluted stream. In the septic zone, for example, the intolerant insects, such as the mayfly nymph, disappear.

---

[213]Metcalf & Eddy, Inc., *Wastewater Engineering: Treatment, Disposal, Reuse.* 3rd. ed. New York: McGraw-Hill, pp. 359–439, 1991.

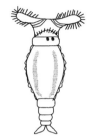

Figure 12.5 Philodina, a common rotifer.

Only air-breathing or specially adapted insects such as mosquito larvae can survive the low dissolved oxygen level in the septic zone. When the stream has completely purified the organic waste, algae returns. Algae are food for higher life organisms such as insects, which in turn serve as food for fish. This is a general biological succession during the self-purification process.

12.5 OXYGEN SAG (DEOXYGENATION)

Earlier in this discussion, biochemical oxygen demand (BOD) was defined as the amount of oxygen required to decay or break down a certain amount of organic matter. Measuring the BOD of a stream is one way to determine how polluted it is. When too much organic waste, such as raw sewage, is added to the stream, all of the available oxygen will be used up. The high BOD reduced the DO because they are interrelated. A typical DO-versus-time-or-distance curve is somewhat spoon-shaped due to the reaeration process. This spoon-shaped curve, commonly called the *oxygen sag curve*, is obtained using the Streeter-Phelps Equation (to be discussed later).

An oxygen sag curve is a graph of the measured concentration of DO in water samples collected upstream from a significant point source (PS) of readily degradable organic material (pollution), from the area of the discharge, and from some distance downstream from the discharge, plotted by sample location. The amount of DO is typically high upstream, diminishes at and just downstream from the discharge location (causing a sag in the line graph), and returns to the upstream levels at some distance downstream from the source of pollution or discharge.

From the oxygen sag curve presented in Figure 12.6, it becomes clear that the percentage of DO versus time or distance shows a characteristic sag that occurs because the organisms breaking down the wastes use up the DO in the decomposition process. When the wastes are decomposed, recovery takes place, and the DO rate rises again.

Several factors determine the extent of recovery. The minimum level of DO found below a sewage outfall depends on the BOD strength and quantity of the

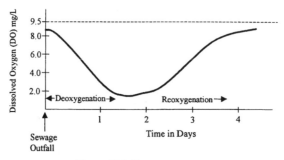

Figure 12.6 Oxygen sag curve.

waste, as well as other factors including velocity of the stream, stream length, biotic content, and the initial DO content.[214]

The rates of reaeration and deoxygenation determine the amount of DO in the stream. If there is no reaeration, the DO will reach zero in a short period of time after the initial discharge of sewage into the stream. But due to reaeration, the rate of which is influenced directly by the rate of deoxygenation, there is enough compensation for aerobic decomposition of organic matter. If the velocity of the stream is too low and the stream is too deep, the DO level may reach zero.

The depletion of oxygen causes a deficit in oxygen, which in turn causes absorption of atmospheric oxygen at the air-liquid interface. Thorough mixing due to turbulence brings about effective reaeration. A shallow, rapid stream will have a higher rate of reaeration (constantly saturated with oxygen) and will purify itself faster than a deep, sluggish one.[215]

✓ *Note:* Reoxygenation of a stream is effected through aeration, absorption, and photosynthesis. Riffles and other natural turbulence in streams enhance aeration and oxygen absorption. Aquatic plants add oxygen to the water through transpiration. Oxygen production from photosynthesis of aquatic plants, primarily blue-green algae, slows or ceases at night, creating a diurnal or daily fluctuation in DO levels in streams. The amount of DO a stream can retain increases as water temperatures cool and concentrations of dissolved solids diminish.

12.6 OTHER FACTORS AFFECTING DO LEVELS IN STREAMS

In the characteristic oxygen sag curve, it is assumed that there is only one point-source discharge of sewage or industrial waste into the stream. The reality is that most streams and rivers have multiple point-source discharge points.

---

[214]Porteous, A., *Dictionary of Environmental Science and Technology* (revised ed.). New York: John Wiley & Sons, Inc., p. 272, 1992.
[215]Smith, R. L., *Ecology and Field Biology*. New York: Harper & Row, p. 223, 1974.

TABLE 12.1. Solubility of Oxygen in Water.

Temperature °C	Solubility mg/L
0	14.6
5	12.8
10	11.3
15	10.1
20	9.2
25	8.4
30	7.6

Note: Values for water exposed to normal air containing 21% oxygen at 760 mm pressure.

A stream can handle the discharges of multiple point-sources if the discharge points are staggered into reaches according to specific lengths based on channel shape, slope, and the composition of the stream bottom. Usually, engineers determine where to place each discharge point.

DO levels in streams can be affected by obstructions in streams that eliminate rapids. Dredging or damming a stream can cause DO levels to drop dramatically. On the other hand, if the dam is high enough to produce turbulence from falling water, DO levels usually return to a high level.

Streams that course their way through forest regions usually contain large amounts of organic matter. Generally emanating from natural sources, these organic deposits are composed of leaves and dead aquatic plants. Decomposition of this organic matter depletes additional DO from the stream by increasing BOD.

Aquatic plants and animals, due to photosynthesis and respiration, cause daily DO fluctuations. During the day, due to photosynthesis, oxygen is produced, which proceeds at the optimum rate during noon hours. At night, on the other hand, consumption of DO by animal organisms occurs during respiration.[216]

The level of DO in a stream is closely linked to temperature. Cooler water results in higher levels of DO. Warmer water results in lower levels of DO. According to Henry's Law, the amount of DO is inversely proportional to the temperature (see Table 12.1). This variation of temperature has a direct influence upon the variety of the fish species found. For example, salmon and trout are species that prefer cooler temperatures.

12.7 IMPACT OF WASTEWATER TREATMENT ON DO LEVELS IN THE STREAM

The dumping of untreated industrial waste or raw sewage (high BOD) into

---

[216]Davis, M.L. and Cornwell, D. A., *Introduction to Environmental Engineering.* New York: McGraw-Hill, p. 173, 1991.

the stream reduces DO levels significantly and can have dramatic impact upon aquatic organisms. In order to reduce the BOD of industrial and sewage waste, wastewater treatment processes are used. Primary wastewater treatment involves passing influent through a screening process, removing grit, and allowing for sedimentation to take place. This process normally reduces BOD by 30–40%. Secondary treatment destroys harmful organisms and removes many dissolved materials. The process is accomplished in two ways.

The trickling filter method passes wastewater over a synthetic media material or crushed stone. The filter media provides a substrate for the growth of a film of microorganisms. The film combines with oxygen and transforms harmful substances into another form that can be filtered out in sedimentation tanks. Another method of secondary treatment is the activated sludge method. This method uses bacteria and oxygen together to destroy harmful microorganisms. Primary and secondary wastewater treatment combined normally reduce BOD by 80–90%.[217]

✓ *Note:* The activated sludge process is analogous to a stream in a box.

12.8 VARIABLES THAT IMPROVE AND DEGRADE STREAM QUALITY

Before moving on to a basic discussion about measuring biochemical oxygen demand (BOD) and dissolved oxygen (DO), a discussion of the variables involved with improving and degrading stream quality is presented. Computer programs that address water pollution can be used to analyze these variables. One such computer program developed by Harmon allows for prediction of the effects of manipulating one or more variables.[218] It should be noted that the particular computer program used in this discussion assumes ideal conditions with variance occurring in specific parameters only.

In the examples shown in Table 12.2, variables such as the type of body of water, temperature, dumping rate, type of waste, waste treatment (if any), and a specific timeframe are listed in the data tables. Three different water body types are featured, ponds, slow rivers (streams), and fast rivers (streams). The specific parameters vary as follows: temperature ranges set at 1°C and 20°C are used; the waste dumping rate is set at either 7 ppm or 14 ppm; the type of waste will be either industrial waste or sewage; wastewater treatment will be indicated by none, primary, or secondary; and the data are based on a fifteen-day period. By comparing the DO and waste content of the three different water bodies under varying conditions, a clearer understanding of water quality improvement and degradation can be gained.

In Table 12.2a, the effects on a pond environment that receives sewage efflu-

---

[217]Spellman, F. R. and Whiting, N. E., *Water Pollution Control Technology: Concepts and Applications.* Rockville, MD, pp. 271–287, 1999.
[218]Harmon, M., *Water Pollution: A Computer Program.* Danbury, CT: EME Corporation, pp. 1–6, 1993.

ent under varying conditions are shown. The first group of examples is shown with temperatures set at 1°C and dumping rate set at 7 ppm, and results are depicted over a fifteen-day period. Example 1 shows a dramatic decline in DO with a rapid increase in organic waste accumulation. The second and third examples have received, respectively, primary and secondary sewage treatment prior to disposal. It is clearly evident from Examples 2 and 3 that sewage treatment makes a difference in DO levels and organic accumulation rates in a stream during a fifteen-day period.

The second group of examples is shown with temperatures set at 20°C and dumping rate set at 14 ppm, and results are depicted over a period of fifteen days. From Example 4, it is evident that DO levels are depleted more rapidly, and that accumulated organic waste increases. Moreover, even with primary or secondary treatment (Examples 5 and 6), there is an appreciable difference in DO level and waste accumulation, compared to the previous examples, due to increased temperature and dumping rate.

Table 12.2b shows the results of industrial waste dumped into the pond used in Table 12.2a. The parameters have been varied; that is, temperature, dumping rate, with wastewater treatment and without wastewater treatment, have been varied. A comparison between the pond receiving sewage in Table 12.2a and the pond receiving industrial waste in Table 12.2b can be made, clarifying the effects of changing parameters.

In Table 12.2c, Examples 1 through 6 demonstrate that when raw sewage is dumped into a stream, biological conditions are sharply changed. Moreover, even sewage treatment effluent, which may be rich in nutrients, can upset environmental stability.

Table 12.2d shows the effects of industrial waste being dumped into a slow stream under varying conditions. Industrial waste is complex. The same stream water can be drawn into an industrial plant for process activities, e.g., cooling water, become contaminated or overheated, and then be dumped back into the stream with or without pretreatment. The typically clean stream water is drawn from its source and then contaminated before being put back into its natural habitat. Unfortunately, the organisms exposed to industrial waste usually pay a high price, which, in the end, affects all organisms in the food chain. Sterile streams and discolored water are mute testimony of this type of surface water pollution.

Table 12.2e depicts sewage dumped into a fast-flowing stream and the resulting effects on environmental conditions. A fast-flowing stream is constantly aerated with oxygen and will purify itself much faster than a slow stream. This phenomenon is clearly evident when one compares the data in Table 12.2e with the data presented in previous tables dealing with slower streams.

In Table 12.2f, the impact of industrial waste on a fast-flowing stream can be observed. Over the years, many tragic fishkills have been documented from streams that have received sudden influxes of highly toxic chemical pollutants.

TABLE 12.2a. Pond Environment (Sewage Effluent).

1) Day	Oxygen	Waste	2) Day	Oxygen	Waste	3) Day	Oxygen	Waste
1	9.60	3.37	1	9.60	3.02	1	9.60	2.74
2	9.50	4.66	2	9.55	3.66	2	9.59	2.87
3	9.06	6.31	3	9.33	4.49	3	9.55	3.03
4	8.05	8.02	4	8.83	5.34	4	9.45	3.20
5	6.35	9.51	5	7.98	6.09	5	9.28	3.35
6	4.05	10.63	6	6.83	6.65	6	9.05	3.46
7	1.42	11.35	7	5.51	7.01	7	8.78	3.53
8	0.00	11.74	8	4.22	7.20	8	8.52	3.57
9	0.00	11.92	9	3.09	7.29	9	8.30	3.59
10	0.00	11.98	10	2.23	7.32	10	8.13	3.60
11	0.00	12.00	11	1.63	7.33	11	8.01	3.60
12	0.00	12.00	12	1.26	7.33	12	7.93	3.60
13	0.00	12.00	13	1.05	7.33	13	7.89	3.60
14	0.00	12.00	14	0.94	7.33	14	7.87	3.60
15	0.00	12.00	15	0.88	7.33	15	7.86	3.60

Pond: 1°C
Sewage
Rate: 7 ppm
Treatment: None

Pond: 1°C
Sewage
Rate: 7 ppm
Treatment: Primary

Pond: 1°C
Sewage
Rate: 7 ppm
Treatment: Secondary

4) Day	Oxygen	Waste	5) Day	Oxygen	Waste	6) Day	Oxygen	Waste
1	4.00	4.07	1	4.00	3.37	1	4.00	2.81
2	3.79	6.66	2	3.90	4.66	2	3.98	3.07
3	2.92	9.96	3	3.46	6.31	3	3.89	3.40
4	0.90	13.37	4	2.45	8.02	4	3.69	3.74
5	0.00	16.36	5	0.75	9.51	5	3.35	4.04
6	0.00	18.60	6	0.00	10.63	6	2.89	4.26
7	0.00	20.03	7	0.00	11.35	7	2.36	4.40
8	0.00	20.81	8	0.00	11.74	8	1.85	4.48
9	0.00	21.16	9	0.00	11.92	9	1.40	4.52
10	0.00	21.29	10	0.00	11.98	10	1.05	4.53
11	0.00	21.33	11	0.00	12.00	11	0.81	4.53
12	0.00	21.33	12	0.00	12.00	12	0.66	4.53
13	0.00	21.33	13	0.00	12.00	13	0.58	4.53
14	0.00	21.33	14	0.00	12.00	14	0.53	4.53
15	0.00	21.33	15	0.00	12.00	15	0.51	4.53

Pond: 20°C
Sewage
Rate: 14 ppm
Treatment: None

Pond: 20°C
Sewage
Rate: 14 ppm
Treatment: Primary

Pond: 20°C
Sewage
Rate: 14 ppm
Treatment: Secondary

Source: Water Pollution Computer Simulation, © EME Corporation, Danbury, CT.

TABLE 12.2b. Pond Environment (Industrial Waste).

1) Day	Oxygen	Waste	2) Day	Oxygen	Waste	3) Day	Oxygen	Waste
1	9.60	3.37	1	9.60	3.02	1	9.60	2.74
2	9.57	4.73	2	9.58	3.70	2	9.60	2.87
3	9.41	6.68	3	9.51	4.67	3	9.58	3.07
4	9.04	9.08	4	9.32	5.87	4	9.54	3.31
5	8.35	11.77	5	8.89	7.22	5	9.48	3.58
6	7.29	14.61	6	8.44	8.64	6	9.37	3.86
7	5.84	17.42	7	7.72	10.04	7	9.22	4.14
8	4.10	20.07	8	6.85	11.37	8	9.05	4.41
9	2.16	22.45	9	5.88	12.56	9	8.86	4.65
10	0.19	24.51	10	4.90	13.59	10	8.66	4.85
11	0.00	26.20	11	3.96	14.43	11	8.47	5.02
12	0.00	27.54	12	3.14	15.10	12	8.31	5.15
13	0.00	28.56	13	2.46	15.61	13	8.17	5.26
14	0.00	29.29	14	1.93	15.98	14	8.07	5.33
15	0.00	29.81	15	1.54	16.24	15	7.99	5.38

Pond: 1°C
Industrial
Rate: 7 ppm
Treatment: None

Pond: 1°C
Industrial
Rate: 7 ppm
Treatment: Primary

Pond: 1°C
Industrial
Rate: 7 ppm
Treatment: Secondary

4) Day	Oxygen	Waste	5) Day	Oxygen	Waste	6) Day	Oxygen	Waste
1	4.00	4.07	1	4.00	3.37	1	4.00	2.81
2	3.93	6.80	2	3.97	4.73	2	3.99	3.08
3	3.63	10.69	3	3.81	6.68	3	3.96	3.47
4	2.89	15.48	4	3.44	9.08	4	3.89	3.95
5	1.51	20.88	5	2.75	11.77	5	3.75	4.49
6	0.00	26.55	6	1.69	14.61	6	3.54	5.06
7	0.00	32.17	7	0.24	17.42	7	3.25	5.62
8	0.00	37.47	8	0.00	20.07	8	2.90	6.15
9	0.00	42.24	9	0.00	22.45	9	2.51	6.62
10	0.00	46.35	10	0.00	24.51	10	2.12	7.03
11	0.00	49.73	11	0.00	26.20	11	1.74	7.37
12	0.00	52.41	12	0.00	27.54	12	1.42	7.64
13	0.00	54.45	13	0.00	28.56	13	1.14	7.84
14	0.00	55.92	14	0.00	29.29	14	0.93	7.99
15	0.00	56.95	15	0.00	29.81	15	0.78	8.10

Pond: 20°C
Industrial
Rate: 14 ppm
Treatment: None

Pond: 20°C
Industrial
Rate: 14 ppm
Treatment: Primary

Pond: 20°C
Industrial
Rate:14 ppm
Treatment: Secondary

Source: Water Pollution Computer Simulation, © EME Corporation, Danbury, CT.

TABLE 12.2c. Slow Stream (Sewage Effluent).

1) Day	Oxygen	Waste	2) Day	Oxygen	Waste	3) Day	Oxygen	Waste
1	13.27	3.37	1	13.27	3.02	1	13.27	2.74
2	13.16	4.66	2	13.21	3.66	2	13.26	2.87
3	12.76	6.31	3	13.01	4.49	3	13.22	3.03
4	11.97	8.02	4	12.62	5.34	4	13.14	3.20
5	10.94	9.51	5	12.10	6.09	5	13.03	3.35
6	9.95	10.63	6	11.61	6.65	6	12.94	3.46
7	9.25	11.35	7	11.26	7.01	7	12.87	3.53
8	8.86	11.74	8	11.06	7.20	8	12.83	3.57
9	8.68	11.92	9	10.98	7.29	9	12.81	3.59
10	8.62	11.98	10	10.94	7.32	10	12.80	3.60
11	8.60	12.00	11	10.94	7.33	11	12.80	3.60
12	8.60	12.00	12	10.93	7.33	12	12.80	3.60
13	8.60	12.00	13	10.93	7.33	13	12.80	3.60
14	8.60	12.00	14	10.93	7.33	14	12.80	3.60
15	8.60	12.00	15	10.93	7.33	15	12.80	3.60

Slow River: 1°C
Sewage
Rate: 7 ppm
Treatment: None

Slow River: 1°C
Sewage
Rate: 7 ppm
Treatment: Primary

Slow River: 1°C
Sewage
Rate: 7 ppm
Treatment: Secondary

4) Day	Oxygen	Waste	5) Day	Oxygen	Waste	6) Day	Oxygen	Waste
1	7.67	4.07	1	7.67	3.37	1	7.67	2.81
2	7.46	6.66	2	7.56	4.66	2	7.65	3.07
3	6.65	9.96	3	7.16	6.31	3	7.57	3.40
4	5.07	13.37	4	6.37	8.02	4	7.41	3.74
5	3.00	16.36	5	5.34	9.51	5	7.20	4.04
6	1.04	18.60	6	4.35	10.63	6	7.00	4.26
7	0.00	20.03	7	3.65	11.35	7	6.86	4.40
8	0.00	20.81	8	3.26	11.74	8	6.79	4.48
9	0.00	21.16	9	3.08	11.92	9	6.75	4.52
10	0.00	21.29	10	3.02	11.98	10	6.74	4.53
11	0.00	21.33	11	3.00	12.00	11	6.73	4.53
12	0.00	21.33	12	3.00	12.00	12	6.73	4.53
13	0.00	21.33	13	3.00	12.00	13	6.73	4.53
14	0.00	21.33	14	3.00	12.00	14	6.73	4.53
15	0.00	21.33	15	3.00	12.00	15	6.73	4.53

Slow Stream: 20°C
Sewage
Rate: 14 ppm
Treatment: None

Slow Stream: 20°C
Sewage
Rate: 14 ppm
Treatment: Primary

Slow Stream: 20°C
Sewage
Rate:14 ppm
Treatment: Secondary

Source: Water Pollution Computer Simulation, © EME Corporation, Danbury, CT.

TABLE 12.2d. Slow Stream (Industrial Waste Effluent)

1) Day	Oxygen	Waste	2) Day	Oxygen	Waste	3) Day	Oxygen	Waste
1	13.27	3.37	1	13.27	3.02	1	13.27	2.74
2	13.23	4.73	2	13.25	3.70	2	13.26	2.87
3	13.09	6.68	3	13.18	4.67	3	13.25	3.07
4	12.80	9.08	4	13.03	5.87	4	13.22	3.31
5	12.35	11.77	5	12.81	7.22	5	13.17	3.58
6	11.81	14.61	6	12.54	8.64	6	13.12	3.86
7	11.25	17.42	7	12.26	10.04	7	13.06	4.14
8	10.72	20.07	8	11.99	11.37	8	13.01	4.41
9	10.24	22.45	9	11.75	12.56	9	12.96	4.65
10	9.83	24.51	10	11.55	13.59	10	12.92	4.85
11	9.49	26.20	11	11.38	14.43	11	12.89	5.02
12	9.23	27.54	12	11.25	15.10	12	12.86	5.15
13	9.02	28.56	13	11.14	15.61	13	12.84	5.26
14	8.87	29.29	14	11.07	15.98	14	12.83	5.33
15	8.77	29.81	15	11.02	16.24	15	12.82	5.38
Slow River: 1°C			Slow River: 1°C			Slow River: 1°C		
Industrial			Industrial			Industrial		
Rate: 7 ppm			Rate: 7 ppm			Rate: 7 ppm		
Treatment: None			Treatment: Primary			Treatment: Secondary		
4) Day	Oxygen	Waste	5) Day	Oxygen	Waste	6) Day	Oxygen	Waste
1	7.67	4.07	1	7.67	3.37	1	7.67	2.81
2	7.60	6.80	2	7.63	4.73	2	7.66	3.08
3	7.32	10.69	3	7.49	6.68	3	7.63	3.47
4	6.73	15.48	4	7.20	9.08	4	7.57	3.95
5	5.83	20.88	5	6.75	11.77	5	7.48	4.49
6	4.75	26.55	6	6.21	14.61	6	7.38	5.06
7	3.63	32.17	7	5.65	17.42	7	7.26	5.62
8	2.57	37.47	8	5.12	20.07	8	7.16	6.15
9	1.62	42.24	9	4.64	22.45	9	7.06	6.62
10	0.80	46.35	10	4.23	24.51	10	6.98	7.03
11	0.12	49.73	11	3.89	26.20	11	6.91	7.37
12	0.00	52.41	12	3.63	27.54	12	6.86	7.64
13	0.00	54.45	13	3.42	28.56	13	6.82	7.84
14	0.00	55.92	14	3.27	29.29	14	6.79	7.99
15	0.00	56.95	15	3.17	29.81	15	6.77	8.10
Slow River: 20°C			Slow River: 20°C			Slow River: 20°C		
Industrial			Industrial			Industrial		
Rate: 14 ppm			Rate: 14 ppm			Rate:14 ppm		
Treatment: None			Treatment: Primary			Treatment: Secondary		

Source: Water Pollution Computer Simulation, © EME Corporation, Danbury, CT.

TABLE 12.2e. Fast Stream (Sewage Effluent).

1) Day	Oxygen	Waste	2) Day	Oxygen	Waste	3) Day	Oxygen	Waste
1	13.60	3.37	1	13.60	3.02	1	13.60	2.74
2	13.50	4.66	2	13.55	3.66	2	13.59	2.87
3	13.11	6.31	3	13.35	4.49	3	13.55	3.03
4	12.41	8.02	4	13.00	5.34	4	13.48	3.20
5	11.59	9.51	5	12.60	6.09	5	13.40	3.35
6	10.92	10.63	6	12.26	6.65	6	13.33	3.46
7	10.49	11.35	7	12.05	7.01	7	13.29	3.53
8	10.26	11.74	8	11.93	7.20	8	13.27	3.57
9	10.15	11.92	9	11.88	7.29	9	13.26	3.59
10	10.11	11.98	10	11.86	7.32	10	13.25	3.60
11	10.10	12.00	11	11.85	7.33	11	13.25	3.60
12	10.10	12.00	12	11.85	7.33	12	13.25	3.60
13	10.10	12.00	13	11.85	7.33	13	13.25	3.60
14	10.10	12.00	14	11.85	7.33	14	13.25	3.60
15	10.10	12.00	15	11.85	7.33	15	13.25	3.60
Fast River: 1°C			Fast River: 1°C			Fast River: 1°C		
Sewage			Sewage			Sewage		
Rate: 7 ppm			Rate: 7 ppm			Rate: 7 ppm		
Treatment: None			Treatment: Primary			Treatment: Secondary		

4) Day	Oxygen	Waste	5) Day	Oxygen	Waste	6) Day	Oxygen	Waste
1	8.00	4.07	1	8.00	3.37	1	8.00	2.81
2	7.79	6.66	2	7.90	4.66	2	7.98	3.07
3	7.02	9.96	3	7.51	6.31	3	7.90	3.40
4	5.62	13.37	4	6.81	8.02	4	7.76	3.74
5	3.99	16.36	5	5.99	9.51	5	7.60	4.04
6	2.64	18.60	6	5.32	10.63	6	7.46	4.26
7	1.78	20.03	7	4.89	11.35	7	7.38	4.40
8	1.31	20.81	8	4.66	11.74	8	7.33	4.48
9	1.10	21.16	9	4.55	11.92	9	7.31	4.52
10	1.03	21.29	10	4.51	11.98	10	7.30	4.53
11	1.00	21.33	11	4.50	12.00	11	7.30	4.53
12	1.00	21.33	12	4.50	12.00	12	7.30	4.53
13	1.00	21.33	13	4.50	12.00	13	7.30	4.53
14	1.00	21.33	14	4.50	12.00	14	7.30	4.53
15	1.00	21.33	15	4.50	12.00	15	7.30	4.53
Fast River: 20°C			Fast River: 20°C			Fast River: 20°C		
Sewage			Sewage			Sewage		
Rate: 14 ppm			Rate: 14 ppm			Rate:14 ppm		
Treatment: None			Treatment: Primary			Treatment: Secondary		

Source: Water Pollution Computer Simulation, © EME Corporation, Danbury, CT.

TABLE 12.2f. Fast Stream (Industrial Waste Effluent).

1) Day	Oxygen	Waste	2) Day	Oxygen	Waste	3) Day	Oxygen	Waste
1	13.60	3.37	1	13.60	3.02	1	13.60	2.74
2	13.57	4.73	2	13.58	3.70	2	13.60	2.87
3	13.43	6.68	3	13.52	4.67	3	13.58	3.07
4	13.17	9.08	4	13.38	5.87	4	13.56	3.31
5	12.80	11.77	5	13.20	7.22	5	13.52	3.58
6	12.39	14.61	6	13.00	8.64	6	13.48	3.86
7	11.99	17.42	7	12.80	10.04	7	13.44	4.14
8	11.61	20.07	8	12.61	11.37	8	13.40	4.41
9	11.27	22.45	9	12.44	12.56	9	13.37	4.65
10	10.98	24.51	10	12.29	13.59	10	13.34	4.85
11	10.74	26.20	11	12.17	14.43	11	13.31	5.02
12	10.55	27.54	12	12.07	15.10	12	13.29	5.15
13	10.40	28.56	13	12.00	15.61	13	13.28	5.26
14	10.30	29.29	14	11.95	15.98	14	13.27	5.33
15	10.22	29.81	15	11.91	16.24	15	13.26	5.38

Fast River: 1°C
Industrial
Rate: 7 ppm
Treatment: None

Fast River: 1°C
Industrial
Rate: 7 ppm
Treatment: Primary

Fast River: 1°C
Industrial
Rate: 7 ppm
Treatment: Secondary

4) Day	Oxygen	Waste	5) Day	Oxygen	Waste	6) Day	Oxygen	Waste
1	8.00	4.07	1	8.00	3.37	1	8.00	2.81
2	7.93	6.80	2	7.97	4.73	2	7.99	3.08
3	7.66	10.69	3	7.83	6.68	3	7.97	3.47
4	7.13	15.48	4	7.57	9.08	4	7.91	3.95
5	6.40	20.88	5	7.20	11.77	5	7.84	4.49
6	5.59	26.55	6	6.79	14.61	6	7.76	5.06
7	4.79	32.17	7	6.39	17.42	7	7.68	5.62
8	4.03	37.47	8	6.01	20.07	8	7.60	6.15
9	3.35	42.24	9	5.67	22.45	9	7.53	6.62
10	2.76	46.35	10	5.38	24.51	10	7.48	7.03
11	2.28	49.73	11	5.14	26.20	11	7.43	7.37
12	1.89	52.41	12	4.95	27.45	12	7.39	7.64
13	1.60	54.45	13	4.80	28.56	13	7.36	7.84
14	1.39	55.92	14	4.70	29.29	14	7.34	7.99
15	1.24	56.95	15	4.62	29.81	15	7.32	8.10

Fast River: 20°C
Industrial
Rate: 14 ppm
Treatment: None

Fast River: 20°C
Industrial
Rate: 14 ppm
Treatment: Primary

Fast River: 20°C
Industrial
Rate:14 ppm
Treatment: Secondary

Source: Water Pollution Computer Simulation, © EME Corporation, Danbury, CT.

12.9 MEASURING BIOCHEMICAL OXYGEN DEMAND (BOD)

The BOD test requires a commitment of five days from initial sample collection (see Chapter 13, "Biological Sampling") to the end of the analysis. During this time, samples are initially seeded with microorganisms and supplied with a carbon nutrient source of glucose-glutamic acid. The sample is then introduced to an environment suitable for bacterial growth at reproducible temperatures, nutrient sources, and light within a 20°C incubator such that oxygen will be consumed. Quality controls, standards, and dilutions are also run for accuracy and precision. Determination of the DO within the samples can be made through Winkler titration. The difference in initial DO readings (prior to incubation) and final DO readings (after a five-day incubation period) predicts the BOD of the sample. A suitable detection limit as per environmental quality control is 1 mg/L^{-1}.[219]

12.9.1 BOD CALCULATIONS

The following steps can be used to calculate BOD and are based on the addition of a nutrient source (carbon-glucose-glutamic acid) and no nutrient source.

(1) The BOD of the blanks (no nutrient source) = DO (final) − DO (initial)
(2) The BOD of the nutrient-added samples = DO (final) − DO (initial) × dilution factor per 300 mL (*Note:* 300 mL is based on the volume contained in BOD bottles.)

The BOD of the sample and standards are calculated by subtracting the final DO from the initial DO and multiplying this factor by the dilution factor. The final value is determined by subtracting the BOD for the blank from the BOD that has been nutrient enriched.

12.10 MEASURING DISSOLVED OXYGEN IN A STREAM[220]

Measurement of DO levels in a stream is accomplished using a dissolved oxygen test kit (e.g., Hach Test Kit). It is recommended (if possible) that water samples be collected at the same time and in the same location as where the temperature measurement was taken.

---

[219]*Standard Methods for the Examination of Water and Waste Water,* 17th ed. Method 507, Washington, DC: American Public Health Association, p. 531, 1985.
[220]Mitchell, M. K. and Stapp, W. B., *Field Manual for Water Quality Monitoring.* Dubuque, IA: Kendall/Hunt Publishers, p. 304, 1996.

12.10.1 USING THE HACH DO TEST KIT

Instructions for using the Hach DO test kit are as follows:

(1) Collect a water sample in a BOD bottle by totally submerging the bottle in the water; remember to stopper the bottle tightly before bringing it to the surface and to make sure there are no air bubbles in the bottle
(2) Add the contents of Hach powder pillows #1 (manganous sulfate) and #2 (alkaline iodide azide) to the bottle; shake the bottle, again making sure there are no air bubbles in it; if oxygen is present in the water, a brownish floc (precipitate) will form
(3) Allow the sample to stand until the precipitate settles halfway; shake the bottle again to see if more floc forms; again wait for the precipitate to settle
(4) Add the contents of powder #3 (sulfamic acid); shake the bottle again, and this time, the floc should dissolve, and the water will turn yellow
(5) Fill the measuring tube (from the kit) with the yellow DO sample; pour the contents into a mixing bottle; pour the second full measuring tube containing the same sample into the mixing bottle; add sodium thiosulfate titrant, one drop at a time, to the sample in the mixing bottle; as the sodium thiosulfate titrant is being added, swirl the sample; count the number of drops added; stop when the color changes from yellow to clear
(6) Divide the number of drops added to the sample by two, which will give you the DO concentration in mg/L^{-1}

Perform the test carefully or the results will not be valid. The results obtained from the analysis will be in milligrams per liter (mg/L^{-1}). Milligrams per liter is the same as parts per million (ppm). Temperature will influence the amount of DO in the water sample. If percent saturation is the desired end result, then convert the mg/L^{-1} to % saturation using Figure 12.7.[221]

As an example, if the water temperature was 12°C and the DO was measured at 10 mg/L^{-1}, the % saturation of oxygen is 78.

12.11 STREAM PURIFICATION: A QUANTITATIVE ANALYSIS[222]

Before sewage is dumped into a stream, it is important to determine the maximum BOD loading for the stream to avoid rendering it septic. The most common method of ultimate wastewater disposal is discharge into a selected body

---

[221] Michaud, J. P., *A Citizen's Guide to Understanding and Monitoring Lakes and Streams*. Publication #94-149. Olympia, WA: Washington State Dept. of Ecology, pp. 1–13, 1994.
[222] Based on materials provided by USEPA, www.epa.gov/ednrmrl/main/abc/s.htm, pp. 1–4, 2000; *Surface Water and Groundwater Pollution: Water Quality in Rivers and Streams*. http://home.ust.hk/irenelo/main/3/3c_body. htm, pp. 1–3, 2000.

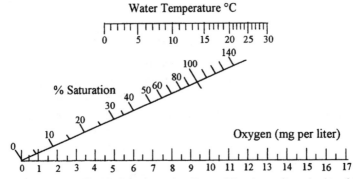

Figure 12.7 Nomogram for calculating saturation. For a quick and easy determination of the percent saturation value for dissolved oxygen at a given temperature, use the saturation chart above. Pair up the mg/L of dissolved oxygen you measured and the temperature of the water in degrees C. Draw a straight line between the water temperature and the mg/L of dissolved oxygen. The percent satruation is the value where the line intercepts the saturation scale. Streams with a saturation value of 90% or above are considered healthy.

of water. The receiving water, stream, lake, or river, is given the final job of purification. The degree of purification that takes place depends on the flow or volume, oxygen content, and reoxygenation ability of the receiving water. Moreover, self-purification is a dynamic variable, changing each day and closely following the hydrological variations characteristic of each stream. Additional variables include stream runoff, water temperature, reaeration, and the time of passage down the stream.

The purification process is carried out by several different aquatic organisms. During the purification of the waste, a sag in the oxygen content of the stream occurs. Mathematical expressions help in determining the oxygen response of the receiving stream. Because the biota and conditions in various parts of the stream change (that is, decomposition of organic matter in a stream is a function of degradation by microorganisms and oxygenation by reaeration, which are competing processes that are working simultaneously), it is difficult to quantify variables and results.

The most common and well-known mathematical equation for oxygen sag for streams and rivers was first described by Streeter and Phelps in 1925. The Streeter-Phelps equation is presented as follows:

$$D = \frac{K_1 L_A}{K_2 - K_1}[e^{-K_1 t} - e^{-K_2 t}] + D_A e^{-K_2 t} \quad (12.1)$$

where

D = Dissolved oxygen deficit (ppm)

t = Time of flow (days)
L_A = Ultimate BOD of the stream after the waste enters
D_A = Initial oxygen deficit (before discharge) (ppm)
K_1 = BOD rate coefficient (per day)
K_2 = Reaeration constant (per day)

✓ *Note:* K_1 or deoxygenation constant is the rate at which microbes consume oxygen for aerobic decomposition of organic matter. The following equation is used to calculate K_1:

$$y = L(1 - 10^{-K_1 t}) \quad \text{or} \quad K_1 = \frac{-\log(1 - y/L)}{5t} \quad (12.2)$$

where:

y = BOD$_5$ (five days BOD)
L = Ultimate or BOD$_{21}$
K_1 = Deoxygenation constant
t = Time in days (five days)

K_2, reaeration constant, is the reaction characteristic of the stream that varies, depending on the velocity of the water, the depth, the surface area exposed to the atmosphere and the amount of biodegradable organic matter in the stream. K_2, or reaeration constants, are given in Table 12.3.

The reaeration constant for a fast-moving, shallow stream is higher than that for a sluggish stream or a lake. The reaeration constant for shallow streams, where vertical gradient and sheer stress exist, is commonly found using the following formulation:

$$K_{2_{20°C}} = \frac{48.6 S^{1/4}}{H^{5/4}} \quad (12.3)$$

TABLE 12.3. Typical Reaeration Constants (K_2) for Water Bodies.

Water Body	Ranges of K_2 at 20°C
Backwaters	0.10–0.23
Sluggish streams	0.23–0.35
Large streams (low velocity)	0.35–0.46
Large streams (normal velocity)	0.46–0.69
Swift streams	0.69–1.15
Rapids	>1.15

The reaeration constant for turbulence that is typical in deep streams can be found using the following equation:

$$K_{2_{20°C}} = \frac{13.0 V^{1/2}}{H^{3/2}} \tag{12.4}$$

where:

K_2 = Reaeration constant
V = Velocity of stream (ft/sec)
H = Stream depth (ft)
S = Slope of stream bed (ft/ft)

The Streeter-Phelps equation should be used with caution because it assumes that conditions such as flow, BOD removal and oxygen demand rate, depth, and temperature throughout the stream are constant. In other words, it assumes that all conditions are the same or constant for every stream; however, this is seldom true. A stream, from reach to reach, changes. Additionally, because rivers and streams are usually longer than they are wide, organic pollution mixes rapidly in these surface waters. Further, some rivers and streams are wider than others. Thus, the mixing of organic pollutants with river or stream water does not occur at the same rate in different rivers and streams.

12.11.1 EXAMPLE 1

Use the following equation and parameters for a stream to calculate the oxygen deficit D in the stream after pollution.

$$D = \frac{K_1 L_A}{K_2 - K_1}[e^{-K_1 t} - e^{-K_2 t}] + D_A e^{-K_2 t}$$

Parameters:

(1) Pollution enters stream at Point X
(2) $t = 2.13$
(3) $L_A = 22$ mg/L (of pollution and stream at Point X)
(4) $D_A = 2$ mg/L
(5) $K_1 = 0.280$/day (base e)
(6) $K_2 = 0.550$/day (base e)

Note: To convert log base e to base 10, divide by 2.31.

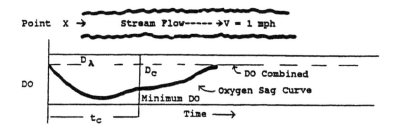

$$D = \frac{0.280 \times 22}{0.550 - 0.280}[e^{-0.280 \times 2.13} - e^{-0.550 \times 2.13}] + 2e^{-0.550 \times 2.13}$$

$$= \frac{6.16}{0.270}[10^{-0.258} - 10^{-0.510}] + 2 \times 10^{-0.510}$$

$$= 22.81[0.5520 - 0.3090] + 2 \times 0.3090$$

$$= 22.81 \times 0.243 + 0.6180 \text{ mg/L}$$

$$= 6.16 \text{ mg/L}$$

12.11.2 EXAMPLE 2

Calculate deoxygenation constant K_1 for a domestic sewage with BOD_5, 135 mg/L and BOD_{21}, 400 mg/L.

$$K_1 = \frac{-\log\left(1 - \dfrac{BOD_1}{BOD_{21}}\right)}{t}$$

$$= \frac{-\log\left(1 - \dfrac{135}{400}\right)}{5}$$

$$= \frac{-\log 0.66}{5}$$

$$= \frac{0.1804}{5}$$

$$= 0.361 / \text{day}$$

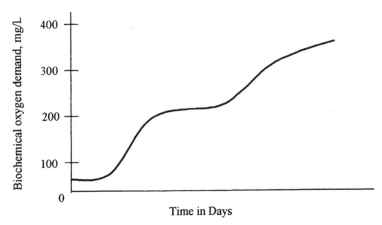

Figure 12.8 Sketch of the carbonaceous and nitrogenous BOD in a waste sample.

It has been determined that BOD occurs in two different phases (see Figure 12.8). The first phase is called the *carbonaceous BOD* (CBOD), because mainly the organic or carbonaceous material is broken down. The second phase is called the *nitrogenous phase*. Here, nitrogen compounds are decomposed, requiring oxygen. This is of particular concern when conducting tests for discharge permit compliance, especially if nitrification is known to occur.

In summary, the Streeter-Phelps equation provides a rough estimate of the ecological conditions in a stream. Small variations in the stream may cause the DO to be higher or lower than the equation indicates. However, this equation may be used in several different ways. The quantity of the waste is important in determining the extent of environmental damage. The Streeter-Phelps equation may be used to determine whether a certain stream has the capability to handle the estimated flow of wastes. In addition, if the stream receives wastes from other sites, then the equation may be helpful in determining whether the stream has recovered fully before it receives the next batch of wastes and, therefore, whether it will be able to recover from the succeeding waste effluents.

12.12 SUMMARY OF KEY TERMS

- In a lentic environment (lake), *oxygen* is added primarily by photosynthetic activity and secondarily by wind-induced wave action.
- In fast streams, *oxygen* is added primarily through reaeration from the atmosphere in rapids, waterfalls, and cascades.
- *DO* concentrations are usually higher and more uniform from surface to bottom in streams than in lakes.
- In the ideal freshwater stream, which has a single point source of pollu-

tion, five zones can be recognized: (1) clean, (2) recent pollution, (3) septic, (4) recovering, and (5) clean.
—The *clean water zone* can be identified by DO concentrations near saturation, low BOD, high species diversity, and the presence of sensitive species.
—In the *recent pollution zone,* DO concentrations are decreasing, BOD is high and increasing, and species diversity is decreasing.
—In the *septic zone,* there is little or no DO, BOD is high but decreasing, and only pollution-tolerant species can be found.
—In the *recovery zone,* DO is increasing to near saturation, BOD is low and decreasing, species diversity is increasing, and sensitive species begin to reappear.
- *Streeter and Phelps* developed a means of estimating the effect of a pollution load on the DO levels in a stream.

12.13 CHAPTER REVIEW QUESTIONS

12.1 The following questions are in regard to various zones within a stream receiving point source pollution.
   (a) Which zone is characterized by high DO and low BOD?

   (b) Low bacterial count and high species diversity is characteristic of which zone?

   (c) High BOD and lower species diversity is characteristic of which zone?

   (d) Blue-green algae and sludgeworms are characteristic of which zone?

   (e) Very little to complete absence of DO with sludgeworms present is characteristic of which zone?

   (f) Number of species increase and number of each species decrease is characteristic of which zone?

12.2 End-of-the-pipe pollution is also known as _____.

12.3 _____ is caused by rainfall or snowmelt moving over and through the ground.

12.4 _____ are organic and inorganic substances that provide food for microorganisms such as bacteria, fungi, and algae.

12.5 The amount of organic matter and the activity by microbial communities living on it is called the _____ of the stream's ecosystem.

12.6 _____ is the amount of oxygen dissolved in a stream.

12.7 We can say that secondary wastewater treatment is analogous to a _____.

12.8 Protozoans are one-celled and _____ on bacteria.

12.9 The _____ is obtained using the Streeter-Phelps Equation.

12.10 Explain Henry's Law.

CHAPTER 13

Biological Sampling

[Rivers] are born traveling, wanting always to move on, intolerant of restraint and interference—itinerant workers always rambling down the line to see what's around the next bend, growling or singing songs, depending on how things suit them. Now, a lake never goes anywhere or does much. It just sort of lies there, slowly dying in the same bed in which it was born. The lake is a set of more or less predictable conditions—at least, compared to the swiftly changing stream of physical, chemical, and biological variables that constitute a living river. Among those variables, though, is one reliable constant—for me, anyway. Whenever I am out on a river some of its freeness rubs off on me. And since freedom is always a highly perishable commodity, frequent returns to the river are necessary for taking on a new supply.—John Madson[223]

13.1 BIOLOGICAL SAMPLING: THE NUTS AND BOLTS OF STREAM ECOLOGY

A few years ago, my sampling partner and I were preparing to perform benthic macroinvertebrate sampling protocols in a wadable section in one of the countless reaches of the Yellowstone River. It was autumn, windy, and cold. Before I stepped into the slow-moving frigid waters, I stood for a moment at the bank and took in the surroundings.

The pallet of autumn is austere in Yellowstone. The coniferous forests east of the Mississippi lack the bronzes, the coppers, the peach-tinted yellows, the livid scarlets that set the mixed stands of the East aflame. All I could see in that line was the quaking aspen and its gold.

This autumnal gold, which provides the closest thing to eastern autumn in the West, is mined from the narrow, rounded crowns of *Populus tremuloides*. The aspen trunks stand stark white and antithetical against the darkness of the

---

[223]Madson, J., *Up on the River.* New York: Lyons Press, pp. 8–15, 1985.

firs and pines, the shiny pale gold leaves sensitive to the slightest rumor of wind. Agitated by the slightest hint of breeze, the gleaming upper surfaces bounced the sun into my eyes. Each tree scintillated, like a shower of gold coins in free fall. The aspens' bright, metallic flash seemed, in all their glittering motion, to make a valiant dying attempt to fill the spectrum of fall.

As bright and glorious as they are, I didn't care that they could not approach the colors of an eastern autumn. While nothing is comparable to experiencing leaf-fall in autumn along the Appalachian Trail, that this autumn was not the same simply didn't matter. This spirited display of gold against dark green lightened my heart and eased the task that was before us, warming the thought of the bone-chilling water and all. With the aspens gleaming gold against the pines and firs, it simply didn't seem to matter.

Notwithstanding the glories of nature alluded to above, one should not be deceived: conducting biological sampling in a stream is not only the nuts and bolts of stream ecology, but it is also very hard and important work.

13.2 BIOLOGICAL SAMPLING: PLANNING

When planning a study that involves biological sampling, it is important to determine the objectives of biological sampling. One important consideration is to determine whether sampling will be accomplished at a single point or at isolated points. Additionally, frequency of sampling must be determined. That is, will sampling be accomplished at hourly, daily, weekly, monthly, or even longer intervals? Whatever sampling frequency is chosen, the entire process will probably continue over a protracted period (i.e., preparing for biological sampling in the field might take several months from the initial planning stages to the time when actual sampling occurs). A stream ecologist should be centrally involved in all aspects of planning.

The USEPA points out that the following issues should be considered in planning the sampling program:[224]

- availability of reference conditions for the chosen area
- appropriate dates to sample in each season
- appropriate sampling gear
- sampling station location
- availability of laboratory facilities
- sample storage
- data management
- appropriate taxonomic keys, metrics, or measurement for macroinvertebrate analysis
- habitat assessment consistency

[224]*Monitoring Water Quality: Intensive Stream Bioassay.* Washington, DC: United States Environmental Protection Agency, pp. 1–35, 08/18/2000; http:www.epa.gov/owow/monitoring/volunteer/stream/vms43.html.

- a USGS topographical map
- familiarity with safety procedures

Once the initial objectives (issues) have been determined and the plan devised, then the sampler can move to other important aspects of the sampling procedure. Along with the items just mentioned, it is imperative that the sampler understand what biological sampling is all about.

Biological sampling allows for rapid and general water quality classification. Rapid classification is possible because quick and easy cross-checking between stream biota and a standard stream biotic index is possible. It is said that biological sampling allows for general water quality classification in the field because sophisticated laboratory apparatus is usually not available. Additionally, stream communities often show a great deal of variation in basic water quality parameters such as DO, BOD, suspended solids, and coliform bacteria. This occurrence can be observed in eutrophic lakes that may vary from oxygen saturation to less than 0.5 mg/L in a single day, and the concentration of suspended solids may double immediately after a heavy rain. Moreover, the sampling method chosen must take into account the differences in the habits and habitats of the aquatic organisms. Tchobanoglous and Schroeder explain that "sampling is one of the most basic and important aspects of water quality management"[225] (again, the nuts and bolts of water quality management).

The first step toward accurate measurement of a stream's water quality is to make sure that the sampling targets those organisms (i.e., macroinvertebrates) that are most likely to provide the information that is being sought.[226] Second, it is essential that representative samples are collected. Laboratory analysis is meaningless if the sample collected was not representative of the aquatic environment being analyzed. As a general rule, samples should be taken at many locations, as often as possible. If, for example, you are studying the effects of sewage discharge into a stream, you should first take at least six samples upstream of the discharge, six samples at the discharge, and at least six samples at several points below the discharge for two to three days (the six-six sampling rule). If these samples show wide variability, then the number of samples should be increased. On the other hand, if the initial samples exhibit little variation, then a reduction in the number of samples may be appropriate.[227]

When planning the biological sampling protocol (using biotic indices as the standards) remember that when the sampling is to be conducted in a stream, findings are based on the presence or absence of certain organisms. Thus, the absence of these organisms must be a function of stream pollution and not of some other ecological problem. The preferred (favored in this text) aquatic

---

[225]Tchobanoglous, G. and Schroeder, E. D., *Water Quality*. Reading, MA: Addison-Wesley, p. 53, 1985.
[226]Mason, C. F., "Biological aspects of freshwater pollution." In *Pollution: Causes, Effects, and Control*. Harrison, R. M. (ed.). Cambridge, Great Britain: The Royal Society of Chemistry, p. 231, 1990.
[227]Kittrell, F. W., *A Practical Guide to Water Quality Studies of Streams*. Washington, DC: U.S. Department of Interior, p. 23, 1969.

group for biological monitoring in streams is the macroinvertebrates, which are usually retained by 30 mesh sieves (pond nets).

13.3 SAMPLING LOCATIONS (STATIONS)

After determining the number of samples to be taken, sampling locations must be determined. Several factors determine where the sampling locations should be set up. These factors include stream habitat types, the position of the wastewater effluent outfalls, the stream characteristics, stream developments (dams, bridges, navigation locks, and other man-made structures), the self-purification characteristics of the stream, and the nature of the objectives of the study.[228]

The stream habitat types used in this discussion are those that are colonized by macroinvertebrates and that generally support the diversity of the macroinvertebrate assemblage in stream ecosystems. Some combination of these habitats would be sampled in a multi-habitat approach to benthic sampling:[229]

- *Cobble (hard substrate)*—cobble is prevalent in the riffles (and runs), which are a common feature throughout most mountain and piedmont streams. In many high-gradient streams, this habitat type will be dominant. However, riffles are not a common feature of most coastal or other low-gradient streams. Sample shallow areas with coarse substrates (mixed gravel, cobble or larger) by holding the bottom of the dip net against the substrate and dislodging organisms by kicking (this is where the "designated kicker," your sampling partner, comes into play) the substrate for 0.5 m upstream of the net.
- *Snags*—snags and other woody debris that have been submerged for a relatively long period (not recent deadfall) provide excellent colonization habitat. Sample submerged woody debris by jabbing in medium-sized snag material (sticks and branches). The snag habitat may be kicked first to help dislodge organisms, but only after placing the net downstream of the snag. Accumulated woody material in pool areas is considered snag habitat. Large logs should be avoided because they are generally difficult to sample adequately.
- *Vegetated banks*—when lower banks are submerged and have roots and emergent plants associated with them, they are sampled in a fashion similar to snags. Submerged areas of undercut banks are good habitats to sample. Sample banks with protruding roots and plants by jabbing

---

[228]Velz, C. J., *Applied Stream Sanitation*. New York: Wiley-Interscience, pp. 313–315, 1970.
[229]Barbour, M. T., Gerritsen, J., Snyder, B. D., and Stribling, J. B., *Revision to Rapid Bioassessment Protocols for Use in Streams and Rivers, Periphyton, Benthic Macroinvertebrates, and Fish*. Washington, DC: United States Environmental Protection Agency, EPA 841-D-97-002, pp. 1–29, 1997; Web site: http://www.epa.gov/owow/monitoring/AWPD/RBP/bioasses.html; USGS, *Field Methods for Hydrologic and Environmental Studies*, Urbana, IL: U.S. Geologic Survey, pp. 1–29, 1999.

into the habitat. Bank habitat can be kicked first to help dislodge organisms, but only after placing the net downstream.
- *Submerged macrophytes*—submerged macrophytes are seasonal in their occurrence and may not be a common feature of many streams, particularly those that are high-gradient. Sample aquatic plants that are rooted on the bottom of the stream in deep water by drawing the net through the vegetation from the bottom to the surface of the water (maximum of 0.5 m each jab). In shallow water, sample by bumping or jabbing the net along the bottom in the rooted area, avoiding sediments where possible.
- *Sand (and other fine sediment)*—usually the least productive macroinvertebrate habitat in streams, this habitat may be the most prevalent in some streams. Sample banks of unvegetated or soft soil by bumping the net along the surface of the substrate rather than dragging the net through soft substrate; this reduces the amount of debris in the sample.

When sampling from a stream for effects of pollution, separate sampling locations should be situated as follows:

> One above the point of receiving; another at the mixing point (approximately 100 feet below discharge); a third location 200 yards down stream; and, the final location should be at least 1 mile downstream. At each location, a number of samples from various spots across the stream should be collected. When sampling downstream of effluent discharges, different sampling arrays may be necessary to obtain truly representative samples.[230]

In a biological sampling program (i.e., based on our experience), the most common sampling methods are the transect and the grid. *Transect sampling* involves taking samples along a straight line either at uniform or at random intervals (see Figure 13.1). The transect involves the cross section of a lake or stream or the longitudinal section of a river or stream. The transect sampling method allows for a more complete analysis by including variations in habitat.

In *grid sampling,* an imaginary grid system is placed over the study area. The grids may be numbered, and random numbers are generated to determine which grids should be sampled (see Figure 13.2). This type of sampling method allows for quantitative analysis because the grids are all of a certain size. For example, to sample a stream for benthic macroinvertebrates, grids that are 0.25 m^2 may be used. Then, the weight or number of benthic macroinvertebrates per square meter can be determined.

Random sampling requires that each possible sampling location have an equal chance of being selected. This can be done by numbering all sampling lo-

---

[230]Hewitt, C. N. and Allott, R., *Understanding Our Environment: An Introduction to Environmental Chemistry and Pollution.* Harrison, R. M. (ed.) Cambridge, Great Britain: The Royal Society of Chemistry, p. 179, 1992.

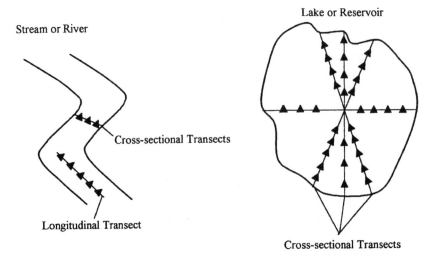

Figure 13.1 Transect sampling.

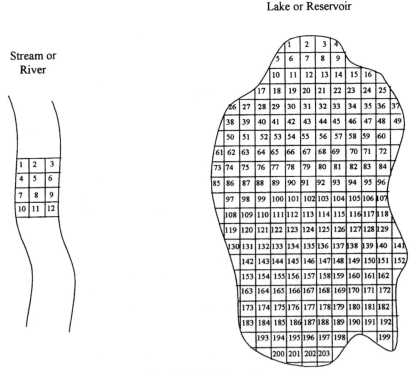

Figure 13.2 Grid sampling

cations, then using a computer, calculator, or a random numbers table to collect a series of random numbers. An illustration of how to put the random numbers to work is provided in the following example. Given a pond that has 300 grid units, find eight random sampling locations using the following sequence of random numbers taken from a standard random numbers table: 101, 209, 007, 018, 099, 100, 017, 069, 096, 033, 041, 011. The first eight numbers of the sequence would be selected and only those grids would be sampled to obtain a random sample.

13.4 STATISTICAL CONCEPTS

Once the samples have been collected and analyzed, it is important to check the accuracy of the results. Probably the most important step in an aquatic study is the *statistical analysis* of the results. The principal concept of statistics is that of variation. In conducting a biological sampling protocol for aquatic organisms, variation is commonly found. Variation comes from the methods that were employed in the sampling process or in the distribution of organisms. Several complex statistical tests can be used to determine the accuracy of data results. In this introductory discussion, however, only basic calculations are presented.

The basic statistical terms include the mean or average, the median, the mode, and the range. The following is an explanation of each of these terms.

(1) *Mean*—is the total of the values of a set of observations divided by the number of observations.

(2) *Median*—is the value of the central item when the data are arrayed in size.

(3) *Mode*—is the observation that occurs with the greatest frequency and thus is the most "fashionable" value.

(4) *Range*—is the difference between the values of the highest and lowest terms.

13.4.1 EXAMPLE 1

Given the following laboratory results for the measurement of dissolved oxygen (DO), find the mean, median, mode, and range.

Data: 6.5 mg/L, 6.4 mg/L, 7.0 mg/L, 6.9 mg/L, 7.0 mg/L

To find the mean:

$$\text{Mean} = \frac{(6.5 \text{ mg/L} + 6.4 \text{ mg/L} + 7.0 \text{ m/L} + 6.0 \text{ mg/L} + 7.0 \text{ mg/L})}{5}$$

$$= 6.58 \text{ mg/L}$$

Mode = 7.0 mg/L (number that appears most often)

Arrange in order: 6.4 mg/L, 6.5 mg/L, 6.9 mg/L, 7.0 mg/L, 7.0 mg/L

Median = 6.9 mg/L (central value)

Range = 7.0 mg/L − 6.4 mg/L = 0.6 mg/L

The importance of using statistically valid sampling methods cannot be overemphasized. Several different methodologies are available. A careful review of the methods available (with the emphasis on designing appropriate sampling procedures) should be made before computing analytic results. Using appropriate sampling procedures along with careful sampling techniques will provide basic data that is accurate.

The need for statistics in environmental sampling is driven by the science itself. Environmental studies often deal with entities that are variable. If there was no variation in environmental data, there would be no need for statistical methods.

Over a given time interval, there will always be some variation in sampling analyses. Usually, the average and the range yield the most useful information. For example, in evaluating the performance of a wastewater treatment plant, a monthly summary of flow measurements, operational data, and laboratory tests for the plant would be used.

In addition to the simple average and range calculations, one may wish to test the precision of the laboratory results. The standard deviation, s, is often used as an indicator of precision. The standard deviation is a measure of the variation (the spread in a set of observations) in the results.

In order to gain a better understanding and perspective on the benefits to be derived from using statistical methods in biological sampling, it is now appropriate to consider some of the basic theory of statistics. In any set of data, the true value (mean) will lie in the middle of all the measurements taken. This is true, providing the sample size is large and only random error is present in the analysis. In addition, the measurements will show a normal distribution, as shown in Figure 13.3.

Figure 13.3 shows that 68.26% of the results fall between $M + s$ and $M − s$, 95.46% of the results lie between $M + 2s$ and $M − 2s$, and 99.74% of the results lie between $M + 3s$ and $M − 3s$. Therefore, if precise, then 68.26% of all the measurements should fall between the true value, estimated by the mean, plus the standard deviation and the true value minus the standard deviation.

Calculation of the sample standard deviation is made using the following formula:

$$s = \sqrt{\frac{\Sigma(X - \bar{X})^2}{n - 1}}$$

where:

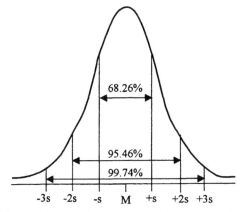

Figure 13.3 The normal distribution curve showing the frequency of a measurement.

s = standard deviation
n = number of samples
X = the measurements from X to X_n
$\overline{X}$ = the mean
Σ = means to sum the values from X to X_n

13.4.2 EXAMPLE 2

Calculate the standard deviation, s, of the following dissolved oxygen values: 9.5, 10.5, 10.1, 9.9, 10.6, 9.5, 11.5, 9.5, 10.0, 9.4

$\overline{X} = 10.0$

X	$X - \overline{X}$	$(X - \overline{X})^2$
9.5	−0.5	0.25
10.5	0.5	0.25
10.1	0.1	0.01
9.9	−0.1	0.01
10.6	0.6	0.36
9.5	−0.5	0.25
11.5	1.5	2.25
9.5	−0.5	0.25
10.0	0	0
9.4	−0.6	0.36
		3.99

$$s = \sqrt{\frac{1}{(10-1)}(3.99)} = \sqrt{\frac{3.99}{9}} = 0.67$$

13.5 SAMPLE COLLECTION[231]

After establishing the sampling methodology and the sampling locations, the frequency of sampling must be determined. The more samples collected, the more reliable the data will be. A frequency of once a week or once a month will be adequate for most aquatic studies. Usually, the sampling period covers an entire year so that yearly variations may be included. The details of sample collection will depend on the type of problem that is being solved and will vary with each study. When a sample is collected, it must be carefully identified with the following information:

- location—name of water body and place of study; longitude and latitude
- date and time
- site—point of sampling (sampling location)
- name of collector
- weather—temperature, precipitation, humidity, wind, etc.
- miscellaneous—any other important information, such as observations
- field notebook—on each sampling day, notes on field conditions should be written. For example, miscellaneous notes and weather conditions can be entered. Additionally, notes that describe the condition of the water are also helpful (color, turbidity, odor, algae, etc.). All unusual findings and conditions should also be entered.

13.5.1 MACROINVERTEBRATE SAMPLING EQUIPMENT

In addition to the appropriate sampling equipment described in Section 13.6, collect the following equipment needed for the macroinvertebrate collection and habitat assessment described in Sections 13.5.2 and 13.5.3.

- jars (two, at least quart size), plastic, wide-mouth with tight cap; one should be empty and the other filled about two-thirds with 70% ethyl alcohol
- hand lens, magnifying glass, or field microscope
- fine-point forceps
- heavy-duty rubber gloves
- plastic sugar scoop or ice-cream scoop
- kick net (rocky-bottom stream) or dip net (muddy-bottom stream)
- buckets (two; see Figure 13.4)
- string or twine (50 yards); tape measure
- stakes (four)

---

[231]From USEPA, *Volunteer Stream Monitoring: A Methods Manual.* Washington, DC: U.S. Environmental Protection Agency, pp. 1–35, 08-18-2000.

Figure 13.4 Sieve bucket. Most professional biological monitoring programs employ sieve buckets as a holding container for composite samples. These buckets have a mesh bottom that allows water to drain while the organisms and debris remain. This material can then be easily transferred to the alcohol-filled jars. However, sieve buckets can be expensive. Many volunteer programs employ alternative equipment, such as the two regular buckets described in this section. Regardless of the equipment, the process for compositing and transferring the sample is basically the same. The decision is one of cost and convenience.

- orange (a stick, an apple, or a fish float may also be used in place of an orange) to measure velocity
- reference maps indicating general information pertinent to the sampling area, including the surrounding roadways, as well as a hand-drawn station map
- station ID tags
- spray water bottle
- pencils (at least 2)

13.5.2 MACROINVERTEBRATE SAMPLING: ROCKY-BOTTOM STREAMS

Rocky-bottom streams are defined as those with bottoms made up of gravel, cobbles, and boulders in any combination. They usually have definite riffle areas. As mentioned, riffle areas are fairly well oxygenated and, therefore, are prime habitats for benthic macroinvertebrates. In these streams, we use the rocky-bottom sampling method described below.

13.5.2.1 Rocky-Bottom Sampling Method

The following method of macroinvertebrate sampling is used in streams that have riffles and gravel/cobble substrates. Three samples are to be collected at each site, and a composite sample is obtained (i.e., one large total sample).

Step 1—A site should have already been located on a map, with its latitude and longitude indicated.

(1) Samples will be taken in three different spots within a 100-yard stream site.

These spots may be three separate riffles; one large riffle with different current velocities; or, if no riffles are present, three run areas with gravel or cobble substrate. Combinations are also possible (if, for example, your site has only one small riffle and several run areas). Mark off the 100-yard stream site. If possible, it should begin at least 50 yards upstream of any human-made modification of the channel, such as a bridge, dam, or pipeline crossing. Avoid walking in the stream, because this might dislodge macroinvertebrates and alter sampling results.

(2) Sketch the 100-yard sampling area. Indicate the location of the three sampling spots on the sketch. Mark the most downstream site as Site 1, the middle site as Site 2, and the upstream site as Site 3.

Step 2—Get into place.

(1) Always approach sampling locations from the downstream end and sample the site farthest downstream first (Site 1). This prevents biasing of the second and third collections with dislodged sediment or macroinvertebrates. Always use a clean kick-seine, relatively free of mud and debris from previous uses. Fill a bucket about one-third full with stream water, and fill your spray bottle.

(2) Select a 3-foot by 3-foot riffle area for sampling at Site 1. One member of the team, the net holder, should position the net at the downstream end of this sampling area. Hold the net handles at a 45 degree angle to the water's surface. Be sure that the bottom of the net fits tightly against the streambed so that no macroinvertebrates escape under the net. You may use rocks from the sampling area to anchor the net against the stream bottom. Do not allow any water to flow over the net.

Step 3—Dislodge the macroinvertebrates.

(1) Pick up any large rocks in the 3-foot by 3-foot sampling area and rub them thoroughly over the partially filled bucket so that any macroinvertebrates clinging to the rocks will be dislodged into the bucket. Then place each cleaned rock outside of the sampling area. After sampling is completed, rocks can be returned to the stretch of stream they came from.

(2) The member of the team designated as the "kicker" should thoroughly stir up the sampling area with their feet, starting at the upstream edge of the 3-foot by 3-foot sampling area and working downstream, moving toward the net. All dislodged organisms will be carried by the stream flow into the net. Be sure to disturb the first few inches of stream sediment to dislodge burrowing organisms. As a guide, disturb the sampling area for about three minutes, or until the area is thoroughly worked over.

(3) Any large rocks used to anchor the net should be thoroughly rubbed into the bucket as above.

Step 4—Remove the net.

(1) Next, remove the net without allowing any of the organisms it contains to wash away. While the net holder grabs the top of the net handles, the kicker grabs the bottom of the net handles and the net's bottom edge. Remove the net from the stream with a forward scooping motion.

(2) Roll the kick net into a cylinder shape and place it vertically in the partially filled bucket. Pour or spray water down the net to flush its contents into the bucket. If necessary, pick debris and organisms from the net by hand. Release back into the stream any fish, amphibians, or reptiles caught in the net.

Step 5—Collect the second and third samples.

(1) Once all of the organisms have been removed from the net, repeat the steps above at Sites 2 and 3. Put the samples from all three sites into the same bucket. Combining the debris and organisms from all three sites into the same bucket is called compositing.

> ✓ *Note:* If your bucket is nearly full of water after you have washed the net clean, let the debris and organisms settle to the bottom. Then, cup the net over the bucket and pour the water through the net into a second bucket. Inspect the water in the second bucket to be sure no organisms came through.

Step 6—Preserve the sample.

(1) After collecting and compositing all three samples, it is time to preserve the sample. All team members should leave the stream and return to a relatively flat section of the stream bank with their equipment. The next step will be to remove large pieces of debris (leaves, twigs, and rocks) from the sample. Carefully remove the debris one piece at a time. While holding the material over the bucket, use the forceps, spray bottle, and your hands to pick, rub, and rinse the leaves, twigs, and rocks to remove any attached organisms. Use a magnifying lens and forceps to find and remove small organisms clinging to the debris. When satisfied that the material is clean, discard it back into the stream.

(2) The water will have to be drained before transferring material to the jar. This process will require two team members. Place the kick net over the second bucket, which has not yet been used and should be completely empty. One team member should push the center of the net into bucket #2, creating a small indentation or depression. Then, hold the sides of the net closely over the mouth of the bucket. The second person can now carefully pour the remaining contents of bucket #1 onto a small area of the net to drain the water and concentrate the organisms. Use care when pouring so that organisms are not lost over the side of the net (see Figure 13.5).

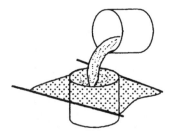

Figure 13.5 Pouring sample water through net.

Use the spray bottle, forceps, sugar scoop, and gloved hands to remove all the material from bucket #1 onto the net. When you are satisfied that bucket #1 is empty, use your hands and the sugar scoop to transfer the material from the net into the empty jar.

Bucket #2 captured the water and any organisms that might have fallen through the netting during pouring. As a final check, repeat the process above, but this time, pour bucket #2 over the net, into bucket #1. Transfer any organisms on the net into the jar.

(3) Now, fill the jar (so that all material is submerged) with the alcohol from the second jar. Put the lid tightly back onto the jar and gently turn the jar upside down two or three times to distribute the alcohol and remove air bubbles.

(4) Complete the sampling station ID tag. Be sure to use a pencil, not a pen, because the ink will run in the alcohol! The tag includes your station number, the stream, location (e.g., upstream from a road crossing), date, time, and the names of the members of the collecting team. Place the ID tag into the sample container, writing side facing out, so that identification can be seen clearly.

13.5.2.2 Rocky-Bottom Habitat Assessment

The habitat assessment (including measuring general characteristics and local land use) for a rocky-bottom stream is conducted in a 100-yard section of stream that includes the riffles from which organisms were collected.

Step 1—Delineate the habitat assessment boundaries.

(1) Begin by identifying the most downstream riffle that was sampled for macroinvertebrates. Using tape measure or twine, mark off a 100-yard section extending 25 yards below the downstream riffle and about 75 yards upstream.

(2) Complete the identifying information on the field data sheet for the habitat

assessment site. On the stream sketch, be as detailed as possible, and be sure to note which riffles were sampled.

Step 2—Describe the General Characteristics and Local Land Use on the field sheet.

(1) For safety reasons as well as to protect the stream habitat, it is best to estimate the following characteristics rather than actually wading into the stream to measure them.
- *Water appearance* can be a physical indicator of water pollution
    —*Clear*—colorless, transparent
    —*Milky*—cloudy-white or gray, not transparent; might be natural or due to pollution
    —*Foamy*—might be natural or due to pollution, generally detergents or nutrients (foam that is several inches high and does not brush apart easily is generally due to pollution)
    —*Turbid*—cloudy brown due to suspended silt or organic material
    —*Dark brown*—might indicate that acids are being released into the stream due to decaying plants
    —*Oily sheen*—multicolored reflection might indicate oil floating in the stream, although some sheens are natural
    —*Orange*—might indicate acid drainage
    —*Green*—might indicate that excess nutrients are being released into the stream
- *Water odor* can be a physical indicator of water pollution
    —*None* or *natural smell*
    —*Sewage*—might indicate the release of human waste material
    —*Chlorine*—might indicate that a sewage treatment plant is over-chlorinating its effluent
    —*Fishy*—might indicate the presence of excessive algal growth or dead fish
    —*Rotten eggs*—might indicate sewage pollution (the presence of a natural gas)
- *Water temperature* can be particularly important for determining whether the stream is suitable as habitat for some species of fish and macroinvertebrates that have distinct temperature requirements. Temperature also has a direct effect on the amount of dissolved oxygen available to aquatic organisms. Measure temperature by submerging a thermometer for at least two minutes in a typical stream run. Repeat once and average the results.
- The *width of the stream channel* can be determined by estimating the width of the streambed that is covered by water from bank to bank. If it varies widely along the stream, estimate an average width.

- *Local land use* refers to the part of the watershed within one-quarter mile upstream of and adjacent to the site. Note which land uses are present, as well as which ones seem to be having a negative impact on the stream. Base observations on what can be seen, what was passed on the way to the stream, and, if possible, what is noticed when leaving the stream.

Step 3—Conduct the habitat assessment.

The following information describes the parameters that will be evaluated for rocky-bottom habitats. Use these definitions when completing the habitat assessment field data sheet.

The first two parameters should be assessed directly at the riffle(s) or run(s) that were used for the macroinvertebrate sampling. The last eight parameters should be assessed in the entire 100-yard section of the stream.

(1) *Attachment sites for macroinvertebrates* are essentially the amount of living space or hard substrates (rocks, snags) available for aquatic insects and snails. Many insects begin their life underwater in streams and need to attach themselves to rocks, logs, branches, or other submerged substrates. The greater the variety and number of available living spaces or attachment sites, the greater the variety of insects in the stream. Optimally, cobble should predominate, and boulders and gravel should be common. The availability of suitable living spaces for macroinvertebrates decreases as cobble becomes less abundant and boulders, gravel, or bedrock become more prevalent.

(2) *Embeddedness* refers to the extent to which rocks (gravel, cobble, and boulders) are surrounded by, covered, or sunken into the silt, sand, or mud of the stream bottom. Generally, as rocks become embedded, fewer living spaces are available to macroinvertebrates and fish for shelter, spawning, and egg incubation.

   ✓ *Note:* To estimate the percent of embeddedness, observe the amount of silt or finer sediments overlaying and surrounding the rocks. If kicking does not dislodge the rocks or cobbles, they might be greatly embedded.

(3) *Shelter for fish* includes the relative quantity and variety of natural structures in the stream, such as fallen trees, logs, and branches; cobble and large rock; and undercut banks that are available to fish for hiding, sleeping, or feeding. A wide variety of submerged structures in the stream provides fish with many living spaces; the more living spaces in a stream, the more types of fish the stream can support.

(4) *Channel alteration* is basically a measure of large-scale changes in the shape of the stream channel. Many streams in urban and agricultural areas have been straightened, deepened (e.g., dredged), or diverted into concrete

channels, often for flood control purposes. Such streams have far fewer natural habitats for fish, macroinvertebrates, and plants than do naturally meandering streams. Channel alteration is present when the stream runs through a concrete channel; when artificial embankments, riprap, and other forms of artificial bank stabilization or structures are present; when the stream is very straight for significant distances; when dams, bridges, and flow-altering structures such as combined sewer overflow (CSO) pipes are present; when the stream is of uniform depth due to dredging; and when other such changes have occurred. Signs that indicate the occurrence of dredging include straightened, deepened, and otherwise uniform stream channels, as well as the removal of streamside vegetation to provide dredging equipment access to the stream.

(5) *Sediment deposition* is a measure of the amount of sediment that has been deposited in the stream channel and the changes to the stream bottom that have occurred as a result of the deposition. High levels of sediment deposition create an unstable and continually changing environment that is unsuitable for many aquatic organisms.

Sediments are naturally deposited in areas where the stream flow is reduced, such as in pools and bends, or where flow is obstructed. These deposits can lead to the formation of islands, shoals, or point bars (sediments that build up in the stream, usually at the beginning of a meander) or can result in the complete filling of pools. To determine whether these sediment deposits are new, look for vegetation growing on them: new sediments will not yet have been colonized by vegetation.

(6) *Stream velocity and depth combinations* are important to the maintenance of healthy aquatic communities. Fast water increases the amount of dissolved oxygen in the water, keeps pools from being filled with sediment; and helps food items like leaves, twigs, and algae move more quickly through the aquatic system. Slow water provides spawning areas for fish and shelters macroinvertebrates that might be washed downstream in higher stream velocities. Similarly, shallow water tends to be more easily aerated (i.e., it holds more oxygen), but deeper water stays cooler longer. Thus, the best stream habitat includes all of the following velocity/depth combinations and can maintain a wide variety of organisms.

slow (<1 ft/sec), shallow (<1.5 ft)
slow, deep
fast, deep
fast, shallow

Measure stream velocity by marking off a 10-foot section of stream run and measuring the time it takes an orange, stick, or other floating biodegradable object to float the 10 feet. Repeat five times, in the same

10-foot section, and determine the average time. Divide the distance (10 feet) by the average time (seconds) to determine the velocity in feet per second.

Measure the stream depth by using a stick of known length and taking readings at various points within your stream site, including riffles, runs, and pools. Compare velocity and depth at various points within the 100-yard site to see how many of the combinations are present.

(7) *Channel flow status* is the percent of the existing channel that is filled with water. The flow status changes as the channel enlarges or as flow decreases as a result of dams and other obstructions, diversions for irrigation, or drought. When water does not cover much of the streambed, the living area for aquatic organisms is limited.

> ✓ *Note:* For the following parameters, evaluate the conditions of the left and right stream banks separately. Define the "left" and "right" banks by standing at the downstream end of the study stretch and look upstream. Each bank is evaluated on a scale of 0–10.

(8) *Bank vegetation protection* measures the amount of the stream bank that is covered by natural (i.e., growing wild and not obviously planted) vegetation. The root system of plants growing on stream banks helps hold soil in place, reducing erosion. Vegetation on banks provides shade for fish and macroinvertebrates and serves as a food source by dropping leaves and other organic matter into the stream. Ideally, a variety of vegetation should be present, including trees, shrubs, and grasses. Vegetation disruption can occur when the grasses and plants on the stream banks are mowed or grazed, or when the trees and shrubs are cut back or cleared.

(9) *Condition of banks* measures erosion potential and whether the stream banks are eroded. Steep banks are more likely to collapse and suffer from erosion than are gently sloping banks and are, therefore, considered to have erosion potential. Signs of erosion include crumbling, unvegetated banks, exposed tree roots, and exposed soil.

(10) The *riparian vegetative zone* is defined as the width of natural vegetation from the edge of the stream bank. The riparian vegetative zone is a buffer zone to pollutants entering a stream from runoff. It also controls erosion and provides stream habitat and nutrient input into the stream.

> ✓ *Note:* A wide, relatively undisturbed riparian vegetative zone reflects a healthy stream system; narrow, far less useful riparian zones occur when roads, parking lots, fields, lawns, and other artificially cultivated areas, bare soil, rocks, or buildings are near the stream bank. The presence of "old fields" (i.e., previously developed agricultural fields allowed to revert to natural conditions) should rate higher than fields in

continuous or periodic use. In arid areas, the riparian vegetative zone can be measured by observing the width of the area dominated by riparian or water-loving plants, such as willows, marsh grasses, and cottonwood trees.

13.5.3 MACROINVERTEBRATE SAMPLING: MUDDY-BOTTOM STREAMS

In muddy-bottom streams, as in rocky-bottom streams, the goal is to sample the most productive habitat available and look for the widest variety of organisms. The most productive habitat is the one that harbors a diverse population of pollution-sensitive macroinvertebrates. Samplers should sample by using a D-frame net (see Figure 13.6) to jab at the habitat and scoop up the organisms that are dislodged. The idea is to collect a total sample that consists of twenty jabs taken from a variety of habitats.

13.5.3.1 Muddy-Bottom Sampling Method

Use the following method of macroinvertebrate sampling in streams that have muddy-bottom substrates.

Step 1—Determine which habitats are present

Muddy-bottom streams usually have four habitats: vegetated banks margins, snags and logs, aquatic vegetation beds and decaying organic matter, and silt/sand/gravel substrate. It is generally best to concentrate sampling efforts on the most productive habitat available, yet to sample other principal habitats if

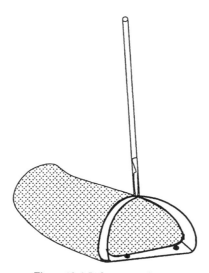

Figure 13.6 D-frame aquatic net.

they are present. This ensures that you will secure as wide a variety of organisms as possible. Not all habitats are present in all streams or are present in significant amounts. If the sampling areas have not been preselected, determine which of the following habitats are present.

✓ *Note:* Avoid standing in the stream while making habitat determinations.

- *Vegetated bank margins* consist of overhanging bank vegetation and submerged root mats attached to banks. The bank margins may also contain submerged, decomposing leaf packs trapped in root wads or lining the streambanks. This is generally a highly productive habitat in a muddy stream, and it is often the most abundant type of habitat.
- *Snags and logs* consist of submerged wood, primarily dead trees, logs, branches, roots, cypress knees, and leaf packs lodged between rocks or logs. This is also a very productive muddy-bottom stream habitat.
- *Aquatic vegetation beds and decaying organic matter* consist of beds of submerged, green/leafy plants that are attached to the stream bottom. This habitat can be as productive as vegetated bank margins and snags and logs.
- *Silt/sand/gravel substrate* includes sandy, silty, or muddy stream bottoms; rocks along the stream bottom; and/or wetted gravel bars. This habitat may also contain algae-covered rocks (*Aufwuchs*). This is the least productive of the four muddy-bottom stream habitats, and it is always present in one form or another (e.g., silt, sand, mud, or gravel might predominate).

Step 2—Determine how many times to jab in each habitat type.

The sampler's goal is to jab a total of 20 times. The D-frame net (see Figure 13.6) is 1 foot wide, and a jab should be approximately 1 foot in length. Thus, 20 jabs equals 20 square feet of combined habitat.

(1) If all four habitats are present in plentiful amounts, jab the vegetated banks 10 times and divide the remaining 10 jabs among the remaining three habitats.
(2) If three habitats are present in plentiful amounts, and one is absent, jab the silt/sand/gravel substrate, the least productive habitat, five times and divide the remaining 15 jabs among the other two more productive habitats.
(3) If only two habitats are present in plentiful amounts, the silt/sand/gravel substrate will most likely be one of those habitats. Jab the silt/sand/gravel substrate five times and the more productive habitat 15 times.
(4) If some habitats are plentiful and others are sparse, sample the sparse habitats to the extent possible, even if you can take only one or two jabs. Take the remaining jabs from the plentiful habitat(s). This rule also applies if you

cannot reach a habitat because of unsafe stream conditions. Jab a total of 20 times.

✓ *Note:* Because the sampler might need to make an educated guess to decide how many jabs to take in each habitat type, it is critical that the sampler note, on the field data sheet, how many jabs were taken in each habitat. This information can be used to help characterize the findings.

Step 3—Get into place.

Outside and downstream of the first sampling location (first habitat), rinse the dip net and check to make sure it does not contain any macroinvertebrates or debris from the last time it was used. Fill a bucket approximately one-third with clean stream water. Also, fill the spray bottle with clean stream water. This bottle will be used to wash the net between jabs and after sampling is completed.

✓ *Note:* This method of sampling requires only one person to disturb the stream habitats. While one person is sampling, a second person should stand outside the sampling area, holding the bucket and spray bottle. After every few jabs, the sampler should hand the net to the second person, who then can rinse the contents of the net into the bucket.

Step 4—Dislodge the macroinvertebrates.

Approach the first sample site from downstream, and sample while walking upstream. Sample in the four habitat types as follows:

(1) Sample vegetated bank margins by jabbing vigorously, with an upward motion, brushing the net against vegetation and roots along the bank. The entire jab motion should occur underwater.
(2) To sample snags and logs, hold the net with one hand under the section of submerged wood being sampled. With the other hand (which should be gloved), rub about 1 square foot of area on the snag or log. Scoop organisms, bark, twigs, or other organic matter dislodged into the net. Each combination of log rubbing and net scooping is one jab.
(3) To sample aquatic vegetation beds, jab vigorously, with an upward motion, against or through the plant bed. The entire jab motion should occur underwater.
(4) To sample a silt/sand/gravel substrate, place the net with one edge against the stream bottom and push it forward about a foot (in an upstream direction) to dislodge the first few inches of silt, sand, gravel, or rocks. To avoid gathering a netful of mud, periodically sweep the mesh bottom of the net back and forth in the water, making sure that water does not run over the top of the net. This will allow fine silt to rinse out of the net. When 20 jabs have been completed, rinse the net thoroughly in the bucket. If necessary,

pick any clinging organisms from the net by hand, and put them in the bucket.

Step 5—Preserve the sample.

(1) Look through the material in the bucket, and immediately return any fish, amphibians, or reptiles to the stream. Carefully remove large pieces of debris (leaves, twigs, and rocks) from the sample. While holding the material over the bucket, use the forceps, spray bottle, and your hands to pick, rub, and rinse the leaves, twigs, and rocks to remove any attached organisms. Use the magnifying lens and forceps to find and remove small organisms clinging to the debris. When satisfied that the material is clean, discard it back into the stream.

(2) Drain the water before transferring material to the jar. This process will require two people. One person should place the net into the second bucket, like a sieve (this bucket, which has not yet been used, should be completely empty) and hold it securely. The second person can now carefully pour the remaining contents of bucket #1 onto the center of the net to drain the water and concentrate the organisms. Use care when pouring so that organisms are not lost over the side of the net. Use the spray bottle, forceps, sugar scoop, and gloved hands to remove all the material from bucket #1 onto the net. When satisfied that bucket #1 is empty, use your hands and the sugar scoop to transfer all the material from the net into the empty jar. The contents of the net can also be emptied directly into the jar by turning the net inside out into the jar. Bucket #2 captured the water and any organisms that might have fallen through the netting. As a final check, repeat the process above, but this time, pour bucket #2 over the net, into bucket #1. Transfer any organisms on the net into the jar.

(3) Fill the jar (so that all material is submerged) with alcohol. Put the lid tightly back onto the jar and gently turn the jar upside down two or three times to distribute the alcohol and remove air bubbles.

(4) Complete the sampling station ID tag (see below). Be sure to use a pencil, not a pen, because the ink will run in the alcohol. The tag should include

```
                STATION ID TAG
    Station #:_____

    Stream:_____

    Location:_____

    Date/Time:_____

    Team Members:_____
    _____
    _____
```

your station number, the stream, location (e.g., upstream from a road crossing), date, time, and the names of the members of the collecting crew. Place the ID tag into the sample container, writing side facing out, so that identification can be seen clearly.

✓ *Note:* To prevent samples from being mixed up, samplers should place the ID tag inside the sample jar.

13.5.3.2 Muddy-Bottom Stream Habitat Assessment

The muddy-bottom stream habitat assessment (which includes measuring general characteristics and local land use) is conducted in a 100-yard section of the stream that includes the habitat areas from which organisms were collected.

✓ *Note:* References made previously, and in the following sections, about a field data sheet (habitat assessment field data sheet) assume that the sampling team is using either the standard forms provided by the USEPA, the USGS, State Water Control Authorities, or generic forms put together by the sampling team. The source of the form and exact type of form are not important. Some type of data recording field sheet should be employed to record pertinent data.

Step 1—Delineate the habitat assessment boundaries.

(1) Begin by identifying the most downstream point that was sampled for macroinvertebrates. Using your tape measure or twine, mark off a 100-yard section extending 25 yards below the downstream sampling point and about 75 yards upstream.
(2) Complete the identifying information on the field data sheet for the habitat assessment site. On the stream sketch, be as detailed as possible, and be sure to note which habitats were sampled.

Step 2—Record General Characteristics and Local Land Use on the data field sheet.

For safety reasons as well as to protect the stream habitat, it is best to estimate these characteristics rather than to actually wade into the stream to measure them. For instructions on completing these sections of the field data sheet, see the rocky-bottom habitat assessment instructions, Section 13.5.2.2.

Step 3—Conduct the habitat assessment.

The following information describes the parameters to be evaluated for muddy-bottom habitats. Use these definitions when completing the habitat assessment field data sheet.

(1) *Shelter for fish and attachment sites for macroinvertebrates* are essentially

the amount of living space and shelter (rocks, snags, and undercut banks) available for fish, insects, and snails. Many insects attach themselves to rocks, logs, branches, or other submerged substrates. Fish can hide or feed in these areas. The greater the variety and number of available shelter sites or attachment sites, the greater the variety of fish and insects in the stream.

> ✓ *Note:* Many of the attachment sites result from debris falling into the stream from the surrounding vegetation. When debris first falls into the water, it is termed new fall, and it has not yet been "broken down" by microbes (conditioned) for macroinvertebrate colonization. Leaf material or debris that is conditioned is called old fall. Leaves that have been in the stream for some time lose their color, turn brown or dull yellow, become soft and supple with age, and might be slimy to the touch. Woody debris becomes blackened or dark in color; smooth bark becomes coarse and partially disintegrated, creating holes and crevices. It might also be slimy to the touch.

(2) *Pool substrate characterization* evaluates the type and condition of bottom substrates found in pools. Pools with firmer sediment types (e.g., gravel, sand) and rooted aquatic plants support a wider variety of organisms than do pools with substrates dominated by mud or bedrock and no plants. In addition, a pool with one uniform substrate type will support far fewer types of organisms than will a pool with a wide variety of substrate types.

(3) *Pool variability* rates the overall mixture of pool types found in the stream according to size and depth. The four basic types of pools are large-yellow, large-deep, small-shallow, and small-deep. A stream with many pool types will support a wide variety of aquatic species. Rivers with low sinuosity (few bends) and monotonous pool characteristics do not have sufficient quantities and types of habitats to support a diverse aquatic community.

(4) *Channel alteration* [see Section 13.5.2.2, Step 3(4)]

(5) *Sediment deposition* [see Section 13.5.2.2, Step 4(5)]

(6) *Channel sinuosity* evaluates the sinuosity or meandering of the stream. Streams that meander provide a variety of habitats (such as pools and runs) and stream velocities and reduce the energy from current surges during storm events. Straight stream segments are characterized by even stream depth and unvarying velocity, and they are prone to flooding. To evaluate this parameter, imagine how much longer the stream would be if it were straightened out.

(7) *Channel flow status* [see Section 13.5.2.2, Step 3(7)]

(8) *Bank vegetative protection* [see Section 13.5.2.2, Step 3(8)]

(9) *Condition of banks* [see Section 13.5.2.2, Step 3(9)]
(10) The *riparian vegetative zone* width [see Section 13.5.2.2, Step 3(10)]

> ✓ *Note:* Whenever stream sampling is to be conducted, it is a good idea to have a reference collection on hand. A reference collection is a sample of locally found macroinvertebrates that have been identified, labeled, and preserved in alcohol. The program advisor, along with a professional biologist/entomologist, should assemble the reference collection, properly identify all samples, preserve them in vials, and label them. This collection may then be used as a training tool and, in the field, as an aid in macroinvertebrate identification.

13.5.4 POST-SAMPLING ROUTINE

After completing the stream characterization and habitat assessment, make sure that all of the field data sheets have been completed properly and that the information is legible. Be sure to include the site's identifying name and the sampling date on each sheet. This information will function as a quality control element.

Before leaving the stream location, make sure that all sampling equipment/devices have been collected and rinsed properly. Double-check to see that sample jars are tightly closed and properly identified. All samples, field sheets, and equipment should be returned to the team leader at this point. Keep a copy of the field data sheet(s) for comparison with future monitoring trips and for personal records.

The next step is to prepare for macroinvertebrate laboratory work. This step includes all the work needed to set up a laboratory for processing samples into subsamples and identifying macroinvertebrates to the family level. A professional biologist/entomologist/stream ecologist or the professional advisor should supervise the identification procedure. [*Note:* The actual laboratory procedures after the sampling and collecting phase are beyond the scope of this text. However, for those who might have interest in follow-up laboratory procedures and relevant information, a list of suggested reading is provided at the end of this chapter.]

13.6 SAMPLING DEVICES

In addition to the standard stream sampling equipment listed in Section 13.5.1, it may be desirable to employ, depending on stream conditions, the use of other sampling devices. Additional sampling devices commonly used, and discussed in the following sections, include dissolved oxygen and temperature monitors, sampling nets (including the D-frame aquatic net), sediment samplers (dredges), plankton samplers, and Secchi disks.

13.6.1 DISSOLVED OXYGEN AND TEMPERATURE MONITOR

As mentioned, the dissolved oxygen (DO) content of a stream sample can provide the investigator with vital information, as DO content reflects the stream's ability to maintain aquatic life. The Winkler DO with Azide Modification Method is commonly used to measure DO content.[232] The Winkler Method can be used in the field but is better suited for laboratory work where better accuracy may be achieved. Basically, the Winkler Method adds a divalent manganese solution followed by a strong alkali to a 300 ml BOD bottle of stream water sample. Any DO rapidly oxidizes an equivalent amount of divalent manganese to basic hydroxides of higher valence states. When the solution is acidified in the presence of iodide, oxidized manganese again reverts to the divalent state, and iodine, equivalent to the original DO content of the sample, is liberated. The amount of iodine is then determined by titration with a standard, usually thiosulfate, solution.

Fortunately for the field biologist, this is the age of miniaturized electronic circuit components and devices; thus, it is not too difficult to obtain portable electronic measuring devices for DO and temperature that are of quality construction and have better than moderate accuracy. These modern electronic devices are usually suitable for laboratory and field use. The device may be subjected to severe abuse in the field; therefore, the instrument must be durable, accurate, and easy to use. Several quality DO monitors are available commercially.

When using a DO monitor, it is important to calibrate (standardize) the meter prior to use. Calibration procedures can be found in *Standard Methods* (latest edition) or in manufacturer's instructions for the meter to be used. Meter calibration is usually accomplished by determining the air temperature, the DO at saturation for that temperature, and then adjusting the meter so that it reads the saturation value. After calibration, the monitor is ready for use. As mentioned, all recorded measurements, including water temperatures and DO readings, should be entered in a field notebook.

13.6.2 SAMPLING NETS

A variety of sampling nets are available for use in the field. The two-person seine net shown in Figure 13.7 is 20 feet long × 4 feet deep with 1/8 inch mesh

---

[232]*Standard Methods for the Examination of Water and Wastwater,* 15th ed. Washington, DC: American Public Health Association, pp. 388–390, 1980.

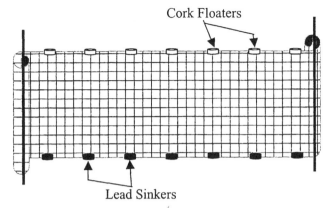

Figure 13.7 Two-person seine net.

and is utilized to collect a variety of organisms. It is used by two people, each holding one end and then walking upstream. Small organisms are easily collected by this method.

Dip nets are used to collect organisms in shallow streams. The surber sampler (collects macroinvertebrates stirred up from the bottom; see Figure 13.8) can be used to obtain a quantitative sample (number of organisms/square feet). It is designed for sampling riffle areas in streams and rivers up to a depth of about 450 mm (18 in.). It consists of two folding stainless steel frames set at right angles to each other. The frame is placed on the bottom, with the net extending downstream. Using your hand or a rake, all sediment enclosed by the frame is dislodged. All organisms are caught in the net and transferred to another vessel for counting.

The D-frame aquatic dip net (see Figure 13.6) is ideal for sweeping over vegetation or for use in shallow streams.

13.6.3 SEDIMENT SAMPLERS (DREDGES)

A sediment sampler or dredge is designed to obtain a sample of the bottom

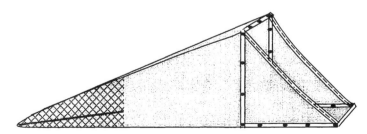

Figure 13.8 Surber sampler.

material in a slow-moving stream and the organisms in it. The simple "homemade" dredge shown in Figure 13.9 works well in water too deep to sample effectively with handheld tools. The homemade dredge is fashioned from a #3 coffee can and a smaller can (see Figure 13.9) with a tight fitting plastic lid (peanut cans work well).

In using the homemade dredge, first, invert it under water so the can fills with water and no air is trapped. Then, lower the dredge as quickly as possible with the "down" line. The idea is to bury the open end of the coffee can in the bottom. Then, quickly pull the "up" line to bring the can to the surface with a minimum loss of material. Dump the contents into a sieve or observation pan to sort. It works best in bottoms composed of sediment, mud, sand, and small gravel.

By using the bottom sampling dredge, a number of different analyses can be made. Because the bottom sediments represent a good area in which to find macroinvertebrates and benthic algae, the communities of organisms living on or in the bottom can be easily studied quantitatively and qualitatively. A chemical analysis of the bottom sediment can be conducted to determine what chemicals are available to organisms living in the bottom habitat.

13.6.4 PLANKTON SAMPLER[233]

Plankton (meaning to drift) are distributed throughout the stream and, in

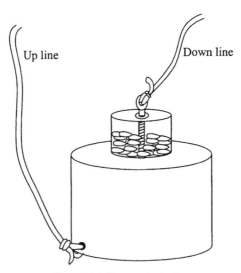

Figure 13.9 Homemade dredge.

---

[233]*Plankton Sampling.* Allendale, MI: Robert B. Annis Water Resource Institute, Grand Valley State University, pp. 1–3, 2000.

Figure 13.10 Plankton net.

particular, in pool areas. They are found at all depths and are comprised of plant (phytoplankton) and animal (zooplankton) forms. Plankton show a distribution pattern that can be associated with the time of day and seasons.

There are three fundamental sizes of plankton: nannoplankton, microplankton, and macroplankton. The smallest are nannoplankton that range in size from 5–60 microns (millionth of a meter). Because of their small size, most nannoplankton will pass through the pores of a standard sampling net. Special fine mesh nets can be used to capture the larger nannoplankton.

Most planktonic organisms fall into the microplankton or net plankton category. The sizes range from the largest nannoplankton to about 2 mm (thousandths of a meter). Nets of various sizes and shapes are used to collect microplankton. The nets collect the organisms by filtering water through fine meshed cloth. The plankton nets on the vessels are used to collect microplankton.

The third group of plankton, as associated with size, is called macroplankton. They are visible to the unaided eye. The largest can be several meters long.

The plankton net or sampler (see Figure 13.10) is a device that makes it possible to collect phytoplankton and zooplankton samples. For quantitative comparisons of different samples, some nets have a flow meter used to determine the amount of water passing through the collecting net.

The plankton net or sampler provides a means of obtaining samples of plankton from various depths so that distribution patterns can be studied. Quantitative determinations can be made by considering the depth of the water column that is sampled. The net can be towed to sample plankton at a single depth (horizontal tow) or lowered into the water to sample the water column (vertical tow). Another possibility is oblique tows where the net is lowered to a predetermined depth and raised at a constant rate as the vessel moves forward.

After towing and removal from the stream, the sides of the net are rinsed to dislodge the collected plankton. If a quantitative sample is desired, a certain quantity of water is collected. If the plankton density is low, then the sample may be concentrated using a low-speed centrifuge or some other filtering device. A definite volume of the sample is studied under the compound microscope for counting and identification of plankton.

13.6.5 SECCHI DISK

For determining water turbidity or degree of visibility in a stream, a Secchi disk is often used (Figure 13.11). The Secchi disk originated with Father Pietro Secchi, an astrophysicist and scientific advisor to the Pope, who was requested to measure transparency in the Mediterranean Sea by the head of the Papal Navy. Secchi used some white disks to measure the clarity of water in the Mediterranean in April of 1865. Various sizes of disks have been used since that time, but the most frequently used disk is an 8-inch diameter metal disk painted in alternate black and white quadrants.

The disk shown in Figure 13.11 is 20 cm in diameter; it is lowered into the stream using the calibrated line. To use the Secchi disk properly, it should be lowered into the stream water until no longer visible. At the point where it is no longer visible, a measurement of the depth is taken. This depth is called the *Secchi disk transparency light extinction coefficient*. The best results are obtained in the lee of a boat or dock. The best readings are usually obtained after early morning and before late afternoon.

13.6.6 MISCELLANEOUS SAMPLING EQUIPMENT

Several other sampling tools/devices are available for use in sampling a stream. For example, consider the standard sand-mud sieve. Generally made of heavy-duty galvanized 1/8″ mesh screen supported by a water-sealed 24 × 15 × 3″ wood frame, this device is useful for collecting burrowing organisms found in soft bottom sediments. Moreover, no stream sampling kit would be complete without a collecting tray, collecting jars of assorted sizes, heavy-duty plastic bags, small pipets, large two-ounce pipets, fine mesh straining net, and black

Figure 13.11 Secchi disk.

china marking pencil. In addition, depending upon the quantity of material to be sampled, it is prudent to include several three- and five-gallon collection buckets in the stream sampling field kit.

13.7 THE BOTTOM LINE: CAN WE AFFORD HEALTHY STREAMS?[234]

The purpose of this discussion has been to present, in a one-volume format, basic information essential to water/wastewater practitioners in furthering their understanding of stream ecology and the self-purification process. The key words mentioned above are *basic information*. This text is like a primer, which sets the scene for more advanced study. [*Note:* For those desiring a more advanced treatment of stream ecology, refer to the suggested reading listed at the end of this chapter.]

Throughout this discussion we have stressed that a stream is a living, dynamic entity. Because it lives, the stream's conditions vary each day and each season; the stream is not an entity that remains static. Moreover, the stream is a living entity that can defend itself against the machinations and pollutants of human beings. As Velz notes, without the stream's ability to self-purify, "the world long ago would have been buried in its own wastes, and water resources would not be fit for any use."[235] Although Velz's statement points out that a

---

[234]Adapted from Karr, J. R. and Chu, E. W., *Restoring Life in Running Waters: Better Biological Monitoring.* Washington, DC: Island Press, p. 174, 1999.
[235]Velz, C. J., *Applied Stream Sanitation.* New York: Wiley-Interscience, p. 8, 1970.

stream can, with time, cleanse itself, it is also true that a stream can become so overloaded with waste that eventually it will suffocate and die. The consequences of such destruction are not only sad to contemplate but are also life-threatening to species that count on the stream for survival.

This discussion also stressed the practice of biological monitoring, employing the use of biotic indices as key measuring tools. We emphasized biotic indices not only for their simplicity of use, but also for the relative accuracy they provide us, although their development and use can sometimes be derailed. The failure of a monitoring protocol to assess environmental condition accurately or to protect running waters usually stems from conceptual, sampling, or analytical pitfalls. Biotic indices can be combined with other tools for measuring the condition of ecological systems in ways that enhance or hinder their effectiveness. The point is, like any other tool, they can be misused. However, that biotic indices can be, and are, misused does not mean that the indices' approach itself is useless.

Thus, to ensure that the biotic indices approach is not useless, it is important for the practicing stream ecologist to remember a few key guidelines:

(1) Sampling everything is not the goal. As Botkin et al. note, biological systems are complex and unstable in space and time, and ecologists often feel compelled to study all components of this variation. Complex sampling programs proliferate. But, every study need not explore everything.[236] Stream ecologists should avoid the temptation to sample all the unique habitats and phenomena that make stream ecology so interesting. Concentration should be placed on the central components of a clearly defined research agenda—detecting and measuring the influence of human activities on the stream's ecological system.

(2) In regard to the influence of human activities on the stream's ecological system, we must see protecting biological condition as a central responsibility of water resource management. One thing is certain, until biological monitoring is seen as essential to track attainment of that goal and biological criteria as enforceable standards mandated by the Clean Water Act, life in the nation's streams will continue to decline.

Humans must take a broader view of the problems in order to devise effective solutions. An increased understanding of science and continued efforts to put pressure on governments to act are needed. Moreover, biological monitoring approaches need to be strengthened. In addition, "the way it's always been

---

[236]Botkin, D. B., *Discordant Harmonies*. New York: Oxford University Press, p. 12, 1990; Pimm, S. L., *The Balance of Nature: Ecological Issues in the Conservation of Species and Communities*. Chicago: University of Chicago Press, p. 32, 1991; Huston, M. A., *Biological Diversity: The Coexistence of Species on Changing Landscapes*. New York: Cambridge University Press, p. 56, 1994; Hilborn, R. and Mangel, M., *The Ecological Detective: Confronting Models with Data*. Princeton, NJ: Princeton University Press, p. 76, 1997.

done" must be reexamined and efforts must be undertaken to do what works to keep streams alive. Agency administrators need to allocate funding inside their own agencies to programs that actually protect stream resources. They should redirect their agencies toward activities they are funding citizen watershed groups to do.

Edmund Muskie asked: "Can we afford healthy waters?"[237] If we answer yes to this question and act now to protect our last best waters and "restore and maintain the physical, chemical, and biological integrity" of the rest, the children of the future might yet find life in running waters. Thus, to Muskie's question, our answer is: We can afford nothing less.

13.8 SUMMARY OF KEY TERMS

- *Kick screen (hand screen)*—is a rectangular piece of screen or netting fastened along two of its edges to vertically held, long handles.
- *Surber sampler*—consists of a sack-like net, framed at the mouth (usually square or rectangular), with an additional frame of the same dimensions extending in front of the mouth at a right angle.
- *Drift net*—is a sack like net, framed at the mouth, and secured in the water so that it can remain for prolonged, unattended periods.
- *Plankton tow*—a fine mesh net that can be pulled through water at various depths; a boat is often required for towing.

13.9 CHAPTER REVIEW QUESTIONS

13.1 Before initiating a sampling study, it is important to determine the _____ of biological sampling.

13.2 List eleven issues that should be considered in planning a sampling program.

13.3 Biological sampling allows for _____ and _____ water quality classification.

13.4 Explain the six-six sampling rule.

---

[237]Muskie, E. S., Testimony of Edmund S. Muskie before the Committee on Environment and Public Work, on the Twentieth Anniversary of Passage of the Clean Water Act. Reprinted as S.Doc. 104-17. *Memorial Tribute Delivered in Congress, Edmund S. Muskie, 1914–1996.* Washington, DC: Government Printing Office, Sept. 1992.

13.5 When planning the biological sampling protocol (using biotic indices as the standard) in a stream, findings are based on the _____ and _____ of certain organisms.

13.6 When lower banks are submerged and have roots and emergent plants associated with them, they are sampled in a fashion similar to _____.

13.7 The _____ sampling method allows for a more complete analysis by including variations in habitat.

13.8 _____ is the value of the central item when the data are arrayed in size.

13.9 The _____ is often used as an indicator of precision.

13.10 _____ are fairly well oxygenated habitats for benthic macroinvertebrates.

13.11 The _____ zone is defined as the width of natural vegetation from the edge of the stream bank.

13.12 _____ streams have four habitats.

13.10 SUGGESTED READING

Brigham, A. R., Brigham, W. U. and Gnilka, A. *Aquatic Insects and Oligochaetes of North and South Carolina.* Mahomet, IL: Midwest Enterprises, 1982.

Cummins, K. W. and Wilzbach, M. A., *Field Procedures for Analysis of Functional Feeding Groups of Stream Macroinvertebrates.* Frostburg, MD: University of Maryland, 1985.

Dates, G. and Byrne, J., *River Watch Network Benthic Macroinvertebrate Monitoring Manual.* Montipelier, VT: River Watch Network, 1995.

Delaware Nature Education Center. *Delaware Stream Watch Guide.* Hockessin, DE: Delaware Nature Society, 1996.

Fore, L., Karr, J. and Wiseman, R., "Assessing invertebrate responses to human activities: evaluating alternative approaches," *Journal of the North American Benthological Society.* 15(2):212–231, 1996.

Hilsenhoff, W. L., *Using a Biotic Index to Evaluate Water Quality in Streams.* Madison, WI: Wisconsin Department of Natural Resources, Technical Bulletin No. 132, 1982.

Hilsenoff, W. L., "Rapid field assessment of organic pollution with a family-level biotic index." *Journal of the North American Benthological Society,* 7:65–68, 1988.

Izaak Walton League of America (IWLA). *A Monitor's Guide to Aquatic Macroinvertebrates.* Gaithersburg, MD: Izaak Walton League of America Save Our Streams, 1992.

Izaak Walton League of America (IWLA). *Stream Insects and Crustaceans Card.* Gaithersburg, MD: Izaak Walton League of America Save Our Streams, no date.

Karr, J. R., "Rivers as sentinels: using the biology of rivers to guide landscape management." In *The Ecology and Management of Streams and Rivers in the Pacific Northwest Coastal Ecoregion.* Springer-Verlag, NY: In press, no date.

Klemm, D. J., et al. *Macroinvertebrate Field and Laboratory Methods for Evaluating the Biological Integrity of Surface Waters.* Cincinnati, OH: U.S. Environmental Protection Agency, EPA/600/4-90/030, 1990.

Lathrop, J., "A Naturalist's Key to Stream Macroinvertebrates for Citizen Monitoring Programs in the Midwest." In *Proceedings of the 1989 Midwest Pollution Control Biologists Meeting,* Chicago Il. Davis, W.S. and Simon, T. P., (ed.), USEPA Region 5 Instream Biocriteria and Ecological Assessment Committee. Chicago, IL: U.S. Environmental Protection Agency, EPA 9059-89/007, 1989.

Maryland Save Our Streams. *Project Heartbeat Volunteer Monitoring Handbook.* Glen Burnie, MD: Maryland Save Our Streams, 1994.

McCafferty, W. P., *Aquatic Entomology: The Fisherman's and Ecologists' Illustrated Guide to Insects and Their Relatives.* Boston: Science Books International, 1981.

McDonald, B., Borden, W. and Lathrop, J., *Citizen Stream Monitoring: A Manual for Illinois.* Illinois Department of Energy and Natural Resources, ILENR/RE-WR90/18, 1990.

Merritt, R. W. and Cummins, K. W. (eds.). *An Introduction to the Aquatic Insects of North America,* 2nd ed. Dubuque, IA: Kendall/Hunt Publishing Company, 1984.

Moen, C. and Schoen, J. "Habitat monitoring." *The Volunteer Monitor* 6(2):1, 1994.

Needham, J. C. and Needham, P. R., *A Guide to the Study of Fresh-Water Biology.* Washington, DC: Reiter's Scientific and Professional Books, 1988.

Peckarsky, B. L. et al., *Freshwater Macroinvertebrates of Northeastern North America.* Ithaca, NY: Cornell University Press, 1990.

Pennak, R. W., *Fresh-Water Invertebrates of the United States: Protozoa to Mollusca,* 3rd ed. New York: John Wiley & Sons, 1989.

Plafkin, J. L., Barbour, M. T., Porter, K. D., Gross, S. K. and Hughes, R. M., *Rapid Bioassessment Protocols for Use in Streams and Rivers: Benthic Macroinvertebrates and Fish.* EPA 440/4-89-001. Washington, DC: U.S. Environmental Protection Agency, 1989.

River Watch Network. *A Simple Picture Key: Major Groups of Benthic Macroinvertebrates Commonly Found in Freshwater New England Streams.* Montpelier, VT: River Watch Network, 1992.

Tennessee Valley Authority (TVA). *Homemade Sampling Equipment.* Water Quality Series Booklet 2. Chattanooga, TN: TVA, 1988.

Tennessee Valley Authority (TVA). *Common Aquatic Flora and Fauna of the Tennessee Valley.* Water Quality Series Booklet 4. Chattanooga, TN: TVA, 1994.

Thorp, J. H. and Covich, A. P. (eds.). *Ecology and Classification of North American Freshwater Invertebrates.* New York: Academic Press, 1991.

USEPA. *Streamwalk Manual.* Seattle, WA: Water Management Divison, U.S. Environmental Protection Agency, 1992.

USEPA. *Biological Criteria: Technical Guidance for Small Streams and Rivers.* EPA 822-B-94-001. Washington, DC: U.S. Environmental Protection Agency, 1994.

USEPA. *The Volunteer Monitor's Guide to Quality Assurance Project Plans.* EPA 841-B-96-003. Washington, DC: U.S. Environmental Protection Agency, 1996.

CHAPTER 14

Final Comprehensive Examination

✓ *Note:* Answers appear in Appendix B.

14.1 A TMDL is essentially a _____.

14.2 _____ refers to the sum of all dissolved constituents in a water sample.

14.3 _____ are net spinners.

14.4 Inhabit the open water limnetic zone of standing waters: _____.

14.5 Most food in a stream comes from _____ the stream.

14.6 Slow-moving streams are dominated by _____ and _____.

14.7 Most streams are primarily _____ food chains.

14.8 A lotic system is _____, while a lentic system is _____.

14.9 Name the three zones of a lotic habitat.

14.10 The major difference between land and freshwater habitat is in the _____ in which they both exist.

14.11 State Hardin's First Law of Ecology.

14.12 What is the main axiom of population ecology?

14.13 Net yield is the same as _____.

14.14 The major ecological unit is:
(a) Biosphere
(b) Lithosphere
(c) Ecosystem
(d) Pond

14.15 Organisms residing within or on the bottom sediments are referred to as:
(a) Benthos
(b) Periphyton
(c) Plankton
(d) Neuston

14.16 Organisms attached to plants or rocks are referred to as:
(a) Benthos
(b) Periphyton
(c) Pelagic
(d) Neuston

14.17 Small plants and animals that move about with the current are:
(a) Plankton
(b) Periphyton
(c) Pelagic
(d) Neuston

14.18 Free-swimming organisms belong to which group of aquatic organisms:
(a) Plankton
(b) Periphyton
(c) Pelagic
(d) Neuston

14.19 Organisms that live on the surface of the water are:
(a) Benthos
(b) Periphyton
(c) Neuston
(d) Pelagic

14.20 Which element is essential in the construction of proteins and amino acids:
   (a) Nitrogen
   (b) Oxygen
   (c) Phosphorus
   (d) Potassium

14.21 The source of phosphorus is:
   (a) Atmosphere
   (b) Hydrosphere
   (c) Lithosphere
   (d) Biosphere

14.22 Rate of reproduction (or birth rate) is called:
   (a) Natality
   (b) Mortality
   (c) Immigration
   (d) Emigration

14.23 Movement of new individuals into a natural area is referred to as:
   (a) Natality
   (b) Mortality
   (c) Immigration
   (d) Emigration

14.24 The maximum number of individuals of a particular species that a natural area can support is:
   (a) Population
   (b) Community
   (c) Ultimate limit
   (d) Carrying capacity

14.25 The main source of nitrogen is:
   (a) Atmosphere
   (b) Hydrosphere
   (c) Lithosphere
   (d) Biosphere

14.26 Most energy of the sun that enters the earth's atmosphere:
   (a) Heats the water
   (b) Is stored as chemical energy
   (c) Causes endothermic reactions
   (d) Raises the earth's temperature

14.27 By what percentage does available energy decrease as it is transferred through the trophic levels:
(a) 40–50%
(b) 60–70%
(c) 80–90%
(d) 90–100%

14.28 Which of the following is not a type of ecological pyramid:
(a) Energy
(b) Number
(c) Productivity
(d) Photosynthesis

14.29 Sunlight, soil, mineral elements, temperature, and moisture are collectively referred to as:
(a) Abiotic factors
(b) Ecosystem
(c) Biotic communities
(d) Autotrophs

14.30 Photosynthesis is the chemical process in which _____ energy is stored as _____ energy:
(a) Solar . . . chemical
(b) Chemical . . . solar
(c) Exothermic . . . kinetic
(d) Cosmic . . . solar

14.31 The combination of organisms (plants and animals) that occupy the same area is referred to as:
(a) Populations
(b) Carrying capacity
(c) Abiotic community
(d) Biotic community

14.32 Which of the following fixes energy of the sun and makes food from simple inorganic substances:
(a) Autotrophs
(b) Heterotrophs
(c) Decomposers
(d) Aufwuchs

14.33 Which of the following is not a heterotroph:
 (a) Herbivore
 (b) Carnivore
 (c) Green plant
 (d) Omnivore

14.34 The _____-carrying capacity is always greater than the _____-carrying capacity.
 (a) Environmental ... ultimate
 (b) Maximum ... ultimate
 (c) Effective ... environmental
 (d) Ultimate ... environmental

14.35 All ecosystems are cyclic mechanisms in which the biotic and abiotic materials are constantly exchanged through:
 (a) Energy webs
 (b) Biogeochemical cycles
 (c) Biomass production
 (d) Population-controlling factors

14.36 The freshwater habitat that is characterized by normally calm water is:
 (a) Lentic
 (b) Lotic
 (c) Littoral
 (d) Rapids

14.37 The region in a lotic environment in which the current is sufficient to prevent sedimentation, providing a firm bottom for organisms, is:
 (a) Littoral zone
 (b) Limnetic zone
 (c) Rapids zone
 (d) Pool zone

14.38 The outermost shallow region of a lentic habitat that has light penetration to the bottom is the:
 (a) Littoral zone
 (b) Limnetic zone
 (c) Rapids zone
 (d) Pool zone

14.39 The region of a lotic habitat where the velocity is reduced and sedimentation occurs is the:
(a) Littoral zone
(b) Limnetic zone
(c) Rapids zone
(d) Pool zone

14.40 The freshwater habitat that is characterized by running waters, e.g., streams, rivers, and springs, is:
(a) Lentic
(b) Lotic
(c) Littoral
(d) Rapids

14.41 Which of the following is not a basic zone in a lentic habitat:
(a) Littoral
(b) Rapids
(c) Limnetic
(d) Profundal

14.42 The lotic habitat can be divided into two basic zones:
(a) Rapids and pools
(b) Littoral and limnetic
(c) Rapids and profundals
(d) Limnetic and pools

14.43 Which of the following uses food stored by the producers, rearranges it, and decomposes some complex material into simple organic compounds:
(a) Autotrophs
(b) Heterotrophs
(c) Decomposers
(d) Aufwuchs

14.44 Density differences that cause stratification in lakes are caused by:
(a) Light intensity
(b) Salinity
(c) Temperature
(d) Photosynthesis

14.45 The upper, usually most oxygenated, layer in a stratified lake is referred to as:
(a) Thermocline
(b) Epilimnion
(c) Hypolimnion
(d) Lentic

14.46 The lowest, usually subject to deoxygenation, layer in a stratified lake is:
(a) Thermocline
(b) Epilimnion
(c) Hypolimnion
(d) Lentic

14.47 Photosynthetic rate depends on:
(a) Dissolved oxygen content of water
(b) Light intensity and photo-period
(c) Stream velocity and depth
(d) Light intensity and salinity

14.48 In a lentic (lake) environment, oxygen is added primarily by reaeration from:
(a) Atmosphere
(b) Rapids
(c) Photosynthetic activity and wind-induced wave action
(d) Respiration

14.49 In a lotic (stream) environment, oxygen is added primarily by:
(a) Reaeration from atmosphere
(b) Photosynthetic activity and wind-induced wave action
(c) Respiration
(d) Runoff

14.50 Nutrients in natural waters that are essential for the synthesis of protoplasm are usually in the form of:
(a) Protoplasm
(b) Biogenic salts
(c) Oxygen
(d) Sewage

14.51 In the natural aging process of a lake, the state at which it is the youngest, has few nutrients, and is characterized by deep and clear water and limited productivity is called:
(a) Eutrophic
(b) Mesotrophic
(c) Oligotrophic
(d) Estuary

14.52 In a lotic environment (streams and rivers), which one of the following is not a primary source of basic nutrients:
(a) Runoff
(b) Dissolution of rocks
(c) Photosynthesis
(d) Sewage

14.53 In a stream receiving sewage discharge, the zone characterized by DO near saturation, high species diversity, and presence of sensitive species is called:
(a) Degradation zone
(b) Recovery zone
(c) Clean zone
(d) Ozone

14.54 The natural phenomenon in lakes occurring in the fall and spring where the water temperature is uniform and a complete mixing of nutrients and oxygen takes place is referred to as:
(a) Stratification
(b) Turnover
(c) Turbidity
(d) Biochemical Oxygen Demand

14.55 Organisms known to prefer a certain set of environmental conditions are known as:
(a) Oligotrophs
(b) Benthic macroinvertebrates
(c) Indicator organisms
(d) Periphyton

14.56 The average amount of oxygen in streams and lakes is:
(a) 4–5 parts per million
(b) 5–6 parts per million
(c) 8–10 parts per million
(d) 11–13 parts per million

14.57 High turbidity can reduce _____, which can limit _____.
 (a) Oxygen solubility . . . Biochemical Oxygen Demand
 (b) Salinity . . . dissolved oxygen
 (c) Dissolved oxygen . . . photosynthesis
 (d) Light penetration . . . photosynthesis

14.58 You must know the stream velocity, depth, and slope of bed to estimate:
 (a) Rate of reaeration
 (b) Photosynthesis
 (c) Deoxygenation constant
 (d) BOD

14.59 A condition or substance that limits the presence and success of an organism or group of organisms is:
 (a) An indicator system
 (b) A limiting factor
 (c) Nutrients
 (d) Species diversity

14.60 The state or condition of a lake being in layers when temperature-induced density produces three distinct layers is:
 (a) Eutrophication
 (b) Reoxygenation
 (c) Deoxygenation
 (d) Stratification

14.61 Interlocked food chains are called a _____.

14.62 The _____ cycle is both sedimentary and gaseous.

14.63 Dissolved oxygen concentrations are usually higher and more uniform from surface to bottom:
 (a) In lakes than in streams
 (b) In streams than in lakes
 (c) In lagoons
 (d) In aeration basins

14.64 The middle layer of a stratified lake, which exhibits a rapidly changing temperature, is known as the:
 (a) Thermocline
 (b) Epilimnion
 (c) Hypolimnion
 (d) Lentic

14.65 A relationship that considers the number of different species and the number of individuals of each species is referred to as:
(a) Indicator organisms
(b) Species diversity
(c) Limiting factors
(d) Density

14.66 The carbon cycle is based on _____.

14.67 When sand particles fall out of the flow, they move by _____.

14.68 The primary source of water to total surface runoff is _____.

14.69 Before initiating a sampling study, it is important to determine the _____ of biological sampling.

14.70 _____ is the amount of oxygen dissolved in a stream.

14.71 Which zone in a point-source polluted stream is characterized by high DO and low BOD?

14.72 Stonefly and mayfly nymphs are _____ to pollution.

14.73 _____ is the process of inventorying aquatic organisms in a selected region of an aquatic system.

14.74 In waters with high turbidity, suspended materials are _____ and light transparency is _____.
(a) Decreased . . . decreased
(b) Decreased . . . increased
(c) Increased . . . decreased
(d) Increased . . . increased

14.75 The term for organic and inorganic substances that provide food for microorganisms such as bacteria, fungi, and algae is:
(a) Dissolved oxygen
(b) Salinity
(c) Eutrophics
(d) Nutrients

14.76 The amount of oxygen dissolved in water and available for organisms is the:
(a) Dissolved oxygen
(b) Dissolved oxygen solubility
(c) BOD
(d) Deoxygenation constant

14.77 Density is defined as:
(a) The amount of suspended solids in a given volume of water
(b) Weight in grams of a given volume of a substance
(c) The weight of a given volume of water (in grams)
(d) The theoretical weight of the substance

14.78 A lake that has high nutrient levels, much undesirable growth, few species but large numbers of each species, and is in the oldest stage in its life cycle is known as:
(a) Mesotrophic
(b) Oligotrophic
(c) Eutrophic
(d) Estuary

14.79 In a stream receiving sewage discharge, the zone characterized by little or no DO, high but decreasing BOD, and only pollution-tolerant organisms is called the:
(a) Recovery zone
(b) Degradation zone
(c) Septic zone
(d) Clean zone

14.80 The natural aging process of all lakes is known as:
(a) Mesotrophication
(b) Eutrophication
(c) Oligotrophication
(d) Hypolimnion

14.81 _____ are often used as indicators of water quality.

14.82 The inactive stage in the metamorphosis of many insects, following the larval stage and preceding the adult form: _____.

14.83 Large wing: _____

14.84 "Darning needles": _____

APPENDIX A

Answers to Chapter Review Questions

Chapter 1 Review Questions

1.1 Ecology
1.2 A major subdivision of ecology in which the individual organism or a species is studied.
1.3 A major subdivision of ecology in which groups of organisms associated together as a unit are studied.
1.4 The source of pollutants that involves discharge of pollutants from an identifiable point.
1.5 An adverse alteration to the environment by a pollutant.
1.6 Temperature; rainfall; light; mineral; wind; humidity; elevation; predominant land forms; tide; medium upon which the organisms exist (water, sand, mud, rock).

Matching:

(1) c
(2) f
(3) n
(4) q
(5) i
(6) t
(7) z
(8) x
(9) k
(10) r
(11) v
(12) a
(13) d
(14) h
(15) o
(16) u
(17) y
(18) w
(19) j
(20) m
(21) p
(22) s
(23) l
(24) b
(25) g
(26) e

237

Label the pond:

A. Producers
B. Primary consumers
C. Secondary consumer
D. Tertiary consumer
E. Decomposers

Chapter 2 Review Questions

2.1 Shifts; coarser; finer
2.2 Precipitation
2.3 A combination process whereby water in plant tissue and in the soil evaporates and transpires to water vapor in the atmosphere.
2.4 Infiltration capacity
2.5 Perennial stream
2.6 They are found at or near the surface and near the center of the stream.
2.7 It is the incorporation of particles when stream velocity exceeds the entraining velocity for a particular particle size.
2.8 Erosion of basin slopes
2.9 Saltation
2.10 Bars
2.11 Thalweg
2.12 It is the bending or curving shape of a stream course.

Chapter 3 Review Questions

3.1 Describes the tendency of chemical elements, including all the essential elements of protoplase, to circulate in the biosphere in characteristic paths from environment to organism and back to the environment.
3.2 Reservoirs
3.3 Residence time
3.4 Mean residence time
3.5 Meteorological, geological, biological
3.6 Local, global
3.7 Gaseous, sedimentary
3.8 Hydrosphere, lithosphere, atmosphere
3.9 Carbon dioxide
3.10 Photosynthesis
3.11 Carbon dioxide; beneficial
3.12 Nitrogen
3.13 Nitrate

Answers to Chapter Review Questions

3.14 Ammonia
3.15 Nitrification
3.16 Denitrification
3.17 Rock
3.18 Fertilizer; algal bloom
3.19 One-problem/one-solution syndrome
3.20 Sulfur

Chapter 4 Review Questions

4.1 Energy may not be created or destroyed.
4.2 Energy; materials
4.3 Flow of energy is from a greater to a lesser amount only.
4.4 Food chain
4.5 Food web
4.6 Decomposers
4.7 It is used to estimate the amount of energy transferred through a food chain.
4.8 It is a graphical representation of the number of organisms at various trophic levels in a food chain, represented by several levels placed one above the other with the base formed by producers, and the apex formed by the final consumer.
4.9 Energy, biomass, numbers
4.10 The energy accumulated by plants
4.11 Net community productivity
4.12 Net primary production

Chapter 5 Review Questions

5.1 The branch of ecology that studies that structure and dynamics of a population.
5.2 Organisms in a population are ecologically equivalent.
5.3 Species; density
5.4 Immigration
5.5 Community ecology
5.6 Clumped distribution
5.7 Carrying capacity
5.8 Environmental carrying capacity
5.9 Population controlling
5.10 Ecological succession
5.11 Bare rocks are exposed to the elements; rocks become colonized by lichens; mosses replace the lichens; grasses and flowering plants replace the

mosses; woody shrubs begin replacing the grasses and flowering plants; a forest eventually grows where bare rocks once existed.
5.12 Pioneer community
5.13 Biotic; abiotic
5.14 We can never do merely one thing. Any intrusion into nature has numerous effects, many of which are unpredictable.

Chapter 6 Review Questions

6.1 300
6.2 3
6.3 The earth neither gains nor loses much matter.
6.4 Humans can live about five minutes without air, about five days without water, and about five weeks without food.
6.5 Use(s)
6.6 Evaporation—evaporation of surface water by the sun's energy
Transpiration—water vapor emitted from plants
Condensation—water changes form from a vapor to liquid
Precipitation—rain, hail, snow, sleet

Chapter 7 Review Questions

7.1 Nature continuously strives to maintain a stream in a clean, healthy, normal state. This is accomplished by maintaining the stream's flora and fauna in a balanced state. Nature balances stream life by maintaining the number and the types of species present in any one part of the stream. Nature structures the stream environment so that interdependency is maintained by a balance between plants and animals.
7.2 Medium
7.3 Twenty
7.4 Lentic; lotic
7.5 Seasonal
7.6 Littoral
7.7 Plankton
7.8 Profundal
7.9 Benthic
7.10 Riffle, run, pool
7.11 Neuston
7.12 A condition or a substance that limits the presence or success of an organism or a group of organisms in an area
7.13 Water quality; temperature; turbidity; DO; acidity; alkalinity; organic or inorganic chemicals; heavy metals; toxic substances; habitat structure; substrate types; water depth and velocity; spatial and temporal complexity

of physical habitats; flow regime; water volume; temporal distribution of flows; energy sources; type, amount, and particle size of organic material entering stream; seasonal pattern of energy availability; biotic interactions; competition; predation; disease; parasitism; mutualism.
7.14 Lowered light penetration
7.15 Carbon dioxide
7.16 Carbon dioxide
7.17 A lake's natural aging process that results in organic material being produced in abundance due to a ready supply of nutrients. It causes a lake to turn into a bog and, eventually, into a terrestrial ecosystem.
7.18 Mesotrophic lake
7.19 Marl
7.20 Open; closed

Chapter 8 Review Questions

8.1 Establishment of a stream; young streams; mature streams; old streams
8.2 Solubility; oxygen
8.3 Eight; ten
8.4 Detritus-based
8.5 Use a metric ruler to measure twice: first the width, then the length. Add the width to the length and divide by two; the result is the average size of the rock.
8.6 Drift
8.7 Current
8.8 Positive rheotaxis
8.9 Genotypic; phenotypic; behavioral; ontogenic
8.10 Accumulate
8.11 Width
8.12 Insects; invertebrates

Chapter 9 Review Questions

9.1 Benthic macroinvertebrates
9.2 Shredders, collectors, scrapers, piercers, predators
9.3 Divers
9.4 Outside
9.5 Family
9.6 (1) The greater the diversity of conditions in a locality, the larger the number of species that make up the community.
(2) The more the conditions deviate from normal, hence from the normal optima of most species, the smaller is the number of species that occur there and the greater the number of individuals of each species.

(3) The longer a locality has been in the same condition, the richer is its biotic community, and the more stable it is.
9.7 Pupae
9.8 Dipteran
9.9 Prolegs
9.10 Riffle beetle

Matching:

(1) h   (14) j
(2) c   (15) g
(3) b   (16) s
(4) t   (17) v
(5) m   (18) y
(6) p   (19) n
(7) i   (20) e
(8) x   (21) k
(9) a   (22) w
(10) z   (23) r
(11) q   (24) o
(12) d   (25) f
(13) u   (26) l

Chapter 10 Review Questions

10.1 To restore and maintain the physical, chemical, and biological integrity of the nation's waters
10.2 Pollution budget
10.3 Bacteria, viruses, protozoa, parasites
10.4 Benthic macroinvertebrates
10.5 Cold water; warm water
10.6 Toxic; organic
10.7 Toxins; toxins; tissues; food chains
10.8 DO, fecal coliform bacteria, nitrate, total phosphate, dissolved solids, suspended sediment
10.9 DO
10.10 Dissolved solids

Chapter 11 Review Questions

11.1 Pollution
11.2 Biomonitoring
11.3 They are ubiquitous; they are relatively sedentary and long-lived; some

species are sensitive to pollution and some are tolerant; they are easy to collect and identify
11.4 Biotic index
11.5 (1) A large dumping of organic waste into a stream tends to restrict the variety of organisms at a certain point in the stream.
(2) As the degree of pollution in a stream increases, key organisms tend to disappear in a predictable order. The disappearance of particular organisms tends to indicate the water quality of the stream.
11.6 Good
11.7 Poor
11.8 Sensitive
11.9 Pollution
11.10 Pollution

Chapter 12 Review Questions

12.1 (a) Clean water zone
    (b) Clean water zone
    (c) Zone of recent pollution
    (d) Zone of recent pollution
    (e) Septic zone
    (f) Recovery zone
12.2 Point source pollution
12.3 Non-point source pollution
12.4 Nutrients
12.5 Saprobity
12.6 Dissolved oxygen
12.7 Stream in a box
12.8 Feed
12.9 Oxygen sag curve
12.10 The amount of DO in water is inversely proportional to the temperature.

Chapter 13 Review Questions

13.1 Objectives
13.2 Availability of reference conditions for the area; appropriate data to sample in each season; appropriate sampling gear; sampling station location; availability of laboratory facilities; sample storage; data management; appropriate taxonomic keys, metrics, or measure for macroinvertebrate analyses; habitat assessment consistency; obtaining a USGS topographical map; and becoming familiar with safety procedures.
13.3 Rapid; general
13.4 When studying the effects of sewage discharge into a stream, you should

first take at least six samples upstream of the discharge, six samples at the discharge, and at least six samples at several points below the discharge for two to three days.
- 13.5 Presence; absence
- 13.6 Snags
- 13.7 Transect
- 13.8 Median
- 13.9 Standard deviation
- 13.10 Riffle areas
- 13.11 Riparian vegetative
- 13.12 Muddy-bottom

APPENDIX B

Answers to Final Comprehensive Examination

14.1 Pollution budget
14.2 Dissolved solids
14.3 Caddisflies
14.4 Planktonic
14.5 Outside
14.6 Insects; invertebrates
14.7 Detritus-based
14.8 Open; closed
14.9 Riffle, run, pool
14.10 Medium
14.11 We can never do merely one thing. Any intrusion into nature has numerous effects, many of which are unpredictable.
14.12 Organisms in a population are ecologically equivalent.
14.13 Net primary production
14.14 c
14.15 a
14.16 b
14.17 a
14.18 c
14.19 c
14.20 a
14.21 c
14.22 a
14.23 c
14.24 c
14.25 a
14.26 d
14.27 c
14.28 d
14.29 a

14.30 a
14.31 d
14.32 a
14.33 c
14.34 d
14.35 b
14.36 a
14.37 c
14.38 a
14.39 d
14.40 b
14.41 b
14.42 a
14.43 b
14.44 c
14.45 b
14.46 c
14.47 b
14.48 c
14.49 a
14.50 b
14.51 c
14.52 c
14.53 c
14.54 b
14.55 c
14.56 c
14.57 d
14.58 a
14.59 b
14.60 d
14.61 Food web
14.62 Sulfur
14.63 b
14.64 a
14.65 b
14.66 Carbon dioxide
14.67 Saltation
14.68 Precipitation
14.69 Objectives
14.70 Dissolved oxygen
14.71 Clean water zone
14.72 Sensitive

14.73 Biomonitoring
14.74 c
14.75 d
14.76 b
14.77 b
14.78 c
14.79 b
14.80 b
14.81 Benthic macroinvertebrates
14.82 Pupae
14.83 Megaloptera
14.84 Dragonflies

Related Reading

Abrahamson, D. E. (ed.)., *The Challenge of Global Warming.* Washington, DC: Island Press, 1988.

Adams, V. D., *Water and Wastewater Examination Manual.* Chelsea, MI: Lewis Publishers, Inc., 1990.

Allen, J. D., *Stream Ecology: Structure and Function of Running Waters.* London: Chapman & Hall, p. 23, 1996.

Arasmith, S., *Introduction to Small Water Systems.* Albany, OR: ACR Publications, Inc., 1993.

Asimov, I., *How Did We Find Out About Photosynthesis?* New York: Walker & Company, 1989.

ASTM. *Manual of Water.* Philadelphia: American Society for Testing and Materials, p. 86, 1969.

Biogeochemical Cycles II: The Nitrogen and Phosphorus Cycles. Yahoo.com internet access, pp. 1–2, 2000.

Barbour, M. T., Gerritsen, J., Snyder, B. D., and Stribling, J. B., *Revision to Rapid Bioassessment Protocols for Use in Streams and Rivers, Periphyton, Benthic Macroinvertebrates and Fish.* Washington, DC: United States Environmental Protection Agency. EPA 841-D-97-002, 1997.

Barlocher, R. and Kendrick, L., "Leaf conditioning by microorganisms." *Oecologia* 20: 359–362, 1975.

Barryman, A. A., *Population Systems: A General Introduction.* New York: Plenum Press, p. 89, 1981.

Benfield, E. F., "Leaf breakdown in streams ecosystems." In *Methods in Stream Ecology.* Hauer, F. R. and Lambertic, G. A. (eds.). San Diego: Academic Press, pp. 579–590, 1996.

Benfield, E. F., Jones, D. R., and Patterson, M. F., "Leaf pack processing in a pastureland stream." *Oikos* 29: 99–103, 1977.

Bolkin, D. E., *Discordant Harmonies.* New York: Oxford University Press, 1990.

Bradbury, I., *The Biosphere.* New York: Helhaven Press, 1991.

Brown, L. R., "Facing food insecurity." In *State of the World.* Brown, L.R. et al. (eds.). New York: W.W. Norton, pp. 179–187, 1994.

Byl, T. D. and Smith, G. F., *Biomonitoring Our Streams: What's It All About?* Nashville, TN: U.S. Geological Survey, pp. 2–3, 1994.

Camann, M., *Freshwater Aquatic Invertebrates: Biomonitoring.* www.humboldt.edu., pp. 1–4, 1996.

Carson, R., *Silent Spring.* Boston: Houghton Mifflin Company, 1962.

Cave, C., *Ecology.* http://web.netcom.com/cristiindex.htm, p. 1, 1998.

Cave, C., *How a River Flows.* http://web.netcome/cristi index.htm/cristi@ix.netcom.com, p. 3, 2000.

Clark, L. R., Gerier, P. W., Hughes, R. D., and Harris, R. F., *The Ecology of Insect Populations.* London: Methuen, p. 73, 1967.

Clutter, F. H., "An empirical biotic index of the quality of water in South African streams and rivers." *Water Resource,* 6: 19–20, 1972.

Conservation group announces "Most endangered" U.S. rivers. In *Water Environment & Technology (WE&T),* p. 13, July 2000.

Cummins, K. W., "Structure and function of stream ecosystems." *Bioscience* 24: 631–641, 1994.

Cummins, K. W. and Klug, M. J., "Feeding ecology of stream invertebrates." *Annual Review of Ecology and Systematics* 10: 631–641, 1979.

Darwin, C., *The Origin of Species,* Suriano, G. (ed.). New York: Grammercy, p. 112, 1998.

Dasmann, R. F., *Environmental Conservation.* New York: John Wiley & Sons, 1984.

Dates, G., *Monitoring Macroinvertebrates.* Presented at USEPA sponsored Fifth National Volunteer Monitoring Conference: Promoting watershed stewardship. Madison, WI: University of Wisconsin, pp. 1–36, August 3–7, 1996.

Davis, M. L. and Cornwell, D. A., *Introduction to Environmental Engineering.* New York: McGraw-Hill, Inc., 1991.

Earle, J. and Callaghan, J., *Impacts of Mine Drainage on Aquatic Life, Water Uses, and Man-made Structures.* Pennsylvania Dept. of Environmental Protection (PA DEP), pp. 1–13, 1998.

Ecology. cristi@ix.netcom.com, pp. 1010, 2000.

Ecosystem Topics, http://www.so.gmu.edu/lrockwoo/ECSSYSTEMS%20 TOPICS.htm, p. 4, 2000.

Einstein, H. A., "Sedimentation (suspended solids)." In *River Ecology and Man.* Oglesby, C., Carlson, A., and McCann, J. (eds.). New York: Academic Press, pp. 309–318, 1972.

Energy Flow through Ecosystems, http//clab.cecil.cc.md.us/faculty/biology/Jason/ef.htm, p. 1, 2000.

Enger, E., Kormelink, J. R., Smith, B. F., and Smith, R. J., *Environmental Science: The Study of Interrelationships.* Dubuque, IA: William C. Brown Publishers, 1989.

Environmental Biology—Ecosystems, http//www.marietta.edu/biol./or/ecosystem.html, p. 6, January 10, 1999.

Evans, R., "Industrial wastes and water supplies." *Journal American Water Works Association* 57: pp. 625–628, 1965.

Evans, E. D. and Neunzig, H. H., "Megaloptera and aquatic nueroptera." In *Aquatic Insects of North America,* 3rd ed. Merritt, R.W. and Cummins, K.W. (eds.). Dubuque, IA: Kendall/Hunt Publishing Company, pp. 298–308, 1996.

Fox, J. C., Fitzgerald, P. R., and Lue-Hing, C., *Sewage Organisms: A Color Atlas.* Chelsea, MI: Lewis Publishers, 1981.

Freedman, B., *Environmental Ecology.* New York: Academic Press, 1989.

Giller, P. S. and Jalmqvist, B., *The Biology of Streams and Rivers.* Oxford: Oxford University Press, 1998.

Gordon, N. D., McMahon, T. A., and Finlayson, B. L., *Stream Hydrology: An Introduction for Ecologists.* Chichester: John Wiley & Sons, Inc., 1992.

Grimaldi, J. V. and Simonds, R. H., *Safety Management.* Homewood, IL: Irwin, 1989.

Halsam, S. M., *River Pollution: An Ecological Perspective.* New York: Belhaven Press, 1990.

Hamburg, M., *Statistical Analysis for Decision Making.* New York: Harcourt Brace Jovanovich, Publishers, 1987.

Harmon, M., *Water Pollution: A Computer Program.* Danbury, CT, EME Corporation, 1993.

Hauer, F. R. and Resh, U. H., "Benthic macroinvertebrates." In *Methods of Stream Ecology.* Hauer, F. R. and Lamberti, G. A. (eds.). San Diego: Academic Press, 1996.

Hem, J. D., *Study and Interpretation of the Chemical Characteristics of Natural Water,* 3rd. ed. Washington, DC: U.S. Geological Survey Water-Supply Paper, p. 263, 1985.

Hewitt, C. N. and Allott, R., *Understanding Our Environment: An Introduction to Environmental Chemistry and Pollution.* R. M. Harrison (ed.). Cambridge, Great Britain: The Royal Society of Chemistry, 1992.

Hiborn, R. and Mangel, M., *The Ecological Detective: Confronting Models with Data.* Princeton, NJ: Princeton University Press, 1997.

Hickman, C. P., Roberts, L. S. and Hickman, F. M., *Integrated Principles of Zoology.* St. Louis: Times Mirror/Mosby College Publishing, 1988.

Hickman, C. P., Roberts, L. S. and Hickman, F. M., *Biology of Animals.* St. Louis: Times Mirror/Mosby College Publishing, 1990.

Hilsenhoff, W. L., *Using a Biotic Index to Evaluate Quality in Streams.* Technical bulletin number 132. Madison, WI, p. 21, 1982.

http: www.epa.gov/owow/monitoring/volunteer/streams/vms 43.html.

http://www.epa.gov/owow//monitoring/awpd/rbp/bioassess.html, 2000.

Huff, W. R., "Biological indices define water quality standards." *Water Environment & Technology,* 5: 21–22, 1993.

Humpesch, U. M., "Effect of temperature on larval growth of *Ecdvonurus dispar* (Ephemeroptera: Heptageriidae) from two English lakes." *Freshwater Biology,* 11, 441–457, 1981.

Huston, M. A., *Biological Diversity: The Coexistence of Species on Changing Landscapes.* New York: Cambridge University Press, 1994.

Hutchinson, G. E., "Thoughts on aquatic insects." *Bioscience,* 32, 495–500, 1981.

Jeffries, M. and Mills, D., *Freshwater Ecology: Principles and Applications.* London: Belhaven Press, 1990.

Karr, C., "Biological integrity and the goal of environmental legislations: lessons for conservation biology." *Conservation Biology* 4:66–84, 1991.

Karr, J. R., *Rivers as Sentinels: Using the Biology of Rivers to Guide Landscape Management.* pacnwfin.mss, revised August 1996.

Karr, J. R. and Chu, E. W., *Restoring Life in Running Waters: Better Biological Monitoring.* Washington, DC: Island Press, 1999.

Karr, J. R. and Dudley, D. R., "Ecological perspective on water quality goals." *Environmental Management* 5: 55–68, 1981.

Kevern, N. R., King, D. L., and Ring, R., "Lake classification systems, part I." *The Michigan Riparian,* p. 1, December 1999.

Killpack, S. C. and Buchholz, D., *Nitrogen in the Environment: Nitrogen.* Missouri: University of Missouri-Columbia, p. 1, 1993.

Kimmel, W. G., "The impact of acid mine drainage on the stream ecosystem." In: *Pennsylvania Coal: Resources, Technology and Utilization.* Majumdar, S.K. and Miller, W.W. (eds.). *The Pa. Acad. Sci. Publ.,* pp. 424–437, 1983.

Kittrell, F. W., *A Practical Guide to Water Quality Studies of Streams.* Washington, DC: U.S. Department of Interior, 1969.

Kolkwitz, R. and Marsson, M., *Okologie der pflanzlichen saprobiean. Berichte der Deutchen botanischen Gesellschaft.*, 26a: pp. 505–519, 1906. (Translated 1976. Biology of plant saprobia), pp. 47–52 in Kemp, L. E., Ingram, W. M., and Mackenthum, K. M., (eds.). *Biology of Water Pollution.* Washington, DC: Federal Water Pollution Control Administration.

Lafferty, P. and Rowe, J., *The Dictionary of Science.* New York: Simon & Schuster, 1993.

Laws, E. A., *An Introductory Text.* New York: John Wiley & Sons, Inc., 1993.

Lenat, D., "Water quality assessment of streams using a qualitative collection method for benthic macroinvertebrates." *Journal North American Benthological Society,* 7: 222–233, 1988.

Lenat, D., "A biotic index for the southeastern United States: Derivation and list of tolerance values, with criteria for assigning water quality ratings." *Journal North American Benthological Society,* 12: 279–290, 1993.

Leopold, L. B., *A View of the River.* Cambridge: MA: Harvard University Press, p. 36, 1994.

Lester, J. N., "Sewage and sewage sludge treatment." *Pollution: Causes, Effects, and Control.* (ed.). Roy M. Harrison. Cambridge, Great Britain: The Royal Society of Chemistry, 1990.

Likens, W.M., "Beyond the shoreline: A watershed ecosystem approach." *Vert. Int. Ver. Theor. AWG. Liminol.* 22, 1–22, 1984.

Mackie, G. L., *Applied Aquatic Ecosystem Concepts.* Ontario, Canada: University of Guelph, 1998.

Madsen, J., *Up On the River.* New York: Lyons Press, pp. 8–15, 1985.

Manual on Water. Philadelphia: American Society for Testing and Materials, 1969.

Maser, C. and Sedell, J. R., *From the Forest to the Sea: The Ecology of Woodland Streams, Rivers, Estuaries, and Oceans.* Delray Beach, FL: St. Lucie Press, p. 37, 1994.

Mason, C. F., "Biological aspects of freshwater pollution." *Pollution: Causes, Effects, and Control.* (ed.) Roy M. Harrison. Cambridge, Great Britain: The Royal Society of Chemistry, 1990.

Masters, G. M., *Introduction to Environmental Engineering and Science.* Englewood Cliffs, NJ: Prentice Hall, 1991.

Mayflies. *Nature.* www.thirteen.org/nature/alienempire/replicators.html. p. 1, September 1, 2000.

McCafferty, P. W., *Aquatic Entomology, the Fishermen's and Ecologists' Illustrated Guide to Insects and Their Relatives.* Boston: Jones and Bartlett Publishers, 1981.

McGhee, T. J., *Water Supply and Sewerage.* New York: McGraw-Hill, Inc., 1991.

Meffe, G. K., "Techno-arrogance and halfway technologies: salmon hatcheries on the Pacific Coast of North America." *Conservation Biology* 6: pp. 350–354, 1992.

Merritt, R. W. and Cummins, K. W., *An Introduction to the Aquatic Insects of North America,* 3rd ed. Dubuque, IA: Kendall/Hunt Publishing Company, p. 1, 1996.

Metcalf & Eddy, Inc., *Wastewater Engineering: Treatment, Disposal, Reuse.* 3rd ed. New York: McGraw-Hill, 1991.

Michaud, J. P., *A Citizen's Guide to Understanding and Monitoring Lakes and Streams.* Publications #94-149. Olympia, Washington State Dept. of Biology, pp. 1–13, 1994.

Miller, G. T., *Environmental Science: An Introduction.* Belmont, CA: Wadsworth, 1988.

Mitchell, M. E. and Stapp, W. B., *Field Manual for Water Quality Monitoring.* Dubuque, IA: Kendall/Hunt Publishers, p. 304, 1996.

Monitoring Water Quality: Intensive Stream Bioassay. Washington, DC: United States Environmental Protection Agency, pp. 1–35, 08/18/2000.

Moran, J. M., Morgan, M. D. and Wiersma, J. H., *Introduction to Environmental Science.* New York: W. H. Freeman and Company, 1986.

Moriswa, M., *Streams: Their Dynamics and Morphology.* New York: McGraw-Hill, p. 66, 1968.

Muskie, E. S., Testimony of Edmund S. Muskie before the Committee on Environment and Public

Work, on the twentieth anniversary of passage of the Clean Water Act. Reprinted as S. Doc. 104-17. *Memorial tribute delivered to Congress, Edmund S. Muskie, 1914–1996.* Washington, DC: Government Printing Office, Sept. 1992.

Naar, J., *Design for a Livable Planet.* New York: Harper & Row, Publishers, 1990.

Narf, R., "Midges, bugs, whirligigs and others: The distribution of insects in lake 'U-Name-It'." *Lakeline N. Am. Lake Manage. Soc.,* 16–17, 57–62, 1997.

National Academy of Sciences, *Eutrophication—Causes, Consequences, Correctives—Proceedings of International Symposium on Eutrophication,* Madison, WI, June 11–15, 1967, Washington, DC: National Academy of Sciences, p. 661, 1969.

Nichols, L. E. and Bulow, F. J., "Effects of acid mine drainage on the stream ecosystem of the East fork of the Obey River, Tennessee." *J. Tenn. Acad. Sci.* v. 48, pp. 30–39, 1973.

Nielson, G. R., *Stoneflies.* Burlington, VT: University of Vermont Extension, pp. 1–3, 1997.

Northington, D. K. and Goodin, J. R., *The Botanical World.* St. Louis: Times Mirror/Mosby College Press, 1984.

O'Connor, D. and Dobbins, W., "The mechanism of reaeration in natural streams." *Journal of Hydraulics Division, Proceedings of American Society of Civil Engineers.* 101: 1315, 1975.

Odum, E. P., *Fundamentals of Ecology.* Philadelphia: Saunders College Publishing, 1971.

Odum, E. P., *Ecology: The Link between the Natural and the Social Sciences.* New York: Holt, Rinehart and Winston, Inc., 1975.

Odum, E. P., *Basic Ecology.* Philadelphia: Saunders College Publishing, 1983.

O'Toole, C., (ed.)., *The Encyclopedia of Insects.* New York: Facts in File, Inc, 1986.

Overcash, M. R. and Davedson, J. M., *Environmental Impact of Nonpoint Source Pollution.* Ann Arbor, MI: Ann Arbor Science Publishers, Inc., 1981.

Parsons, J. D., "Literature pertaining to formation of acid mine waters and their effects on the chemistry and fauna of streams." *Trans. Ill. Stat Acad. Sci.,* v. 50, pp. 49–52, 1957.

Parsons, J. D., "The effects of acid strip-mine effluents on the ecology of a stream." *Arch. Hydrobiol.,* v. 65, pp. 25–50, 1968.

Paul, R. W., Jr., Benfield, E. F. and Cairns, J., Jr., "Effects of thermal discharge on leaf decomposition in a river ecosystem." *Verhandlugen der Internationalen Vereinigung fur Theoretsche and Angewandte Limnologie* 20: 1759–1766, 1978.

Penneck, R. W., *Fresh-water Invertebrates of the United States,* 3rd ed., p. 189, 1989.

Peters, N. E., *Evaluation of Environmental Factors Affecting Yields of Major Dissolved Ions of Streams in the United States.* Washington, DC: U.S. Geological Survey Water-Supply Paper, p. 39, 1984.

Peterson, R. C. and Cummins, K. W., "Leaf processing in a woodland stream." *Freshwater Biology* 4:345–368, 1974.

Phosphorus Cycle. Britannica.com Inc., p. 1, 2000.

Pianka, E. R., *Evolutionary Ecology.* New York: Harper Collins Publisher, 1988.

Pimm, S. L., *The Balance of Nature: Ecological Issues in the Conservation of Species and Communities.* Chicago: University of Chicago Press, 1991.

Plafkin, J. L., Barbour, M. T., Porter, K. D., Gross, S. K. and Hughes, R. M., *Rapid Bioassessment Protocols for Use in Streams and Rivers: Benthic Macroinvertebrates and Fish.* USEPA, Office of Water. EPA/44-4-89-001, 1989.

Plankton Sampling. Allendale, MI: Robert B. Annis Water Resource Institute, Grand Valley State University, 2000.

Pope, P. E., *Forestry and Water Utility: Pollution Control Practices.* West Lafayette, IN: Purdue University, pp. 1–8, 2000.

Porteous, A., *Dictionary of Environmental Science and Technology.* New York: John Wiley & Sons, Inc., 1992.

Price, P. W., *Insect Ecology.* New York: John Wiley & Sons, Inc., 1984.

Richards, K., *Rivers: Form and Processes in Alluvial Channels.* London: Methuen, 1982.

Roback, S. S. and Richardson, J. W., "The effects of acid mine drainage on aquatic insects." *Proc. Acad. Nat. Sci. Phil.* 121: 81–107, 1969.

Rosenburg, D. M., "A National Aquatic ecosystem health program for Canada: We should go against the flow." *Bull. Entomolo. Soc. Can.,* Winnipeg: Freshwater Institute, 30(4): 144–152, 1998.

Sharov, A. A., "Life-system approach: A system paradigm in population ecology." *Oikos* 63: 485–494, 1992.

Sharov, A., *What is Population Ecology?* Blacksburg, VA: Department of Entomology, Virginia Tech. University, pp. 1–2, 1996.

Sharov, A., *Population Ecology.* http://www.gpysymoth.ent.vt.edu/sharov/population/welcome.html. p. 1, 1997.

Smith, R. A., Alexander, R. B. and Wolman, M. G., "Water-quality trends in the Nation's rivers." *Science,* 235(4796): 1607–1615, 1987.

Smith, R. L., *Ecology and Field Biology.* New York: Harper & Row, Publishers, 1974.

Spellman, F. R., *The Science of Water: Concepts and Applications.* Lancaster, PA: Technomic Publishing Company, Inc., pp. 175–176, 1998.

Spellman, F. R., *The Science of Environmental Pollution.* Lancaster, PA: Technomic Publishing Company, Inc., p. 4–5, 1999.

Spellman, F. R., *Microbiology for Water/Wastewater Operators.* Revised ed. Lancaster, PA: Technomic Publishing Company, Inc., pp. 92–93, 2000.

Spellman, F. R. and Whiting, N., *Environmental Science and Technology.* Rockville, MD: Government Institutes, pp. 265–267, 1999.

Standard Methods for Examination of Water and Wastewater. 15th ed. Washington, DC: APHA (American Public Health Association), 1980.

Stewart, K. W. and Stark, B. P., *Nymphs of North American Stonefly General (Plecoptera),* Volume 12. Denton: Thomas Say Foundation, 1988.

Suberkoop, K., Godshalk, G. L., and Klug, M. J., "Changes in the chemical composition of leaves during processing in a woodland stream." *Ecology* 57: 720–727, 1976.

Surface Water and Groundwater Pollution: Water Quality in Rivers and Streams. http://home.ust.hk//irenelo/main/3/3c_body.htm, 2000.

Surface Water: The Teachers Guide. Alexandria, VA: Water Environment Federation, 1988.

Sweeney, B. W. and Vannote, R. L., "Population synchrony in mayflies: A predator satiation hypothesis." *Evolution,* 36, 810–821, 1982.

Tansley, A. G., "The use and abuse of vegetational concepts and terms." *Ecology,* 16: 284–307.

Tchobanoglous, G. and Schroeder, E. D., *Water Supply.* Reading, MA: Addison-Wesley Publishing Company, 1985.

The Many Habitats a Stream Provides. cristi@ix.netcom.com, pp. 1–3, 2000.

Tomera, A. N., *Understanding Basic Ecological Concepts.* Portland, ME: J. Weston Walch, Publisher, 1989.

Trautman, M. B., "Fish distribution and abundance correlated with stream gradients as a consideration in stocking programs." *North American Wildlife Conference,* 7: 283, 1942.

"U.S. EPA signs TMDL Rule despite congressional protests." *Water Environment & Technology (WE&T),* August 2000.

USEPA. *Quality Criteria for Water.* Washington, DC: U.S. Environmental Protection Agency, p. 156, 1976.

USEPA. *Quality Criteria for Water 1986.* Washington, DC: U.S. Environmental Protection Agency, Office of Water, EPA 440/5-86-00 (variously paginated), 1986.

USEPA. *Biological Criteria: National Program Guidance for Surface Waters.* Washington, DC: Environmental Protection Agency, EPA 440-5-90-004, 1990.

USEPA. *What is Nonpoint Source Pollution?* Washington, DC: United States Environmental Protection Agency, EPA-F-94-005, pp. 1–5, 1994.

USEPA. *Biological Assessment Methods, Biocriteria, and Biological Indicators: Bibliography of Selected Technical, Policy, and Regulatory Literature.* Washington, DC: Environmental Protection Agency, EPA 230-B-96-001, 1996.

USEPA. *Monitoring Water Quality: Chapter 4 Macroinvertebrates and Habitats.* www.epa.gov/owow/monitoring/volunteer/stream/ums40.html, 1996.

USEPA. *Summary of State Biological Assessment Programs for Streams and Rivers.* Washington, DC: Environmental Protection Agency, EPA 230-R-96-07, 1996.

USEPA. *Biological Indicators of Watershed Health: Invertebrates as Indicators.* www.epa.gov/seishome/atlas/bioindicators/invertsasindicators. html, pp. 1–2, 1999.

USEPA. *Lotic Ecosystems of the Streambed, ECOVIEW.* Washington, DC: United States Environmental Protection Agency, pp. 1–2, 1999.

USEPA. *Total Maximum Daily Load (TMDL) Program.* Washington, DC: U.S. Environmental Protection Agency, EPA 841-F-00-008, July 2000.

USEPA. *Volunteer Stream Monitoring: A Methods Manual.* Washington, DC: U.S. Environmental Protection Agency, 08-18-2000.

USEPA. www.epa.gov//ednrmrl/main/abc/s.htm, 2000.

USA Today. "Pollution is top environmental concern," August 29, 2000.

USGS, *Field Methods for Hydrological and Environmental Studies,* Urbana, IL: U.S. Geologic Survey, 1999.

USGS, *Hawaiian Volcano Observatory,* http://hvo.wr.usgs.gov/volcano-watch/1999/99_0/_21.html, p. 1.3, February 1999.

USGS, Smith, R. A., Alexander, R. B., and Lanfear, K. J., *Stream Water Quality in the Conterminous United States—Status and Trends of Selected Indicators During the 1980s.* Washington, DC: U.S. Geological Survey (USGS) Water-Supply Paper 2400, February 1997.

Velz, C. J., *Applied Stream Sanitation.* New York: Wiley-Interscience, 1970.

Wackernagel, M., *Framing the Sustainability Crisis: Getting from Concerns to Action,* http://www.sdri.ubc.ca/ publications/ wacherna.html., 1997.

Warner, R. W., "Distribution of biota in a stream polluted by acid mine-drainage." *Ohio J. Sci.,* v. 71, pp. 202–215, 1971.

Water Pollution Control Federation. *Surface Water: The Student's Resource Guide.* Alexandria, VA: WPCF, 1988.

WCED, World Commission on Environment and Development. *Our Common Future.* New York: Oxford University Press, p. 27, 1987.

Weber, S. (Ed.). *USA by Numbers.* Washington, DC: Zero Population Growth, 1988.

Weed, C. E. and Ruschky, C. W., "Benthic macroinvertebrate community structure in a stream receiving acid mine drainage." *Proc. Pa. Acad. Sci.,* 50: 41–46, 1971.

Welch, P. S., *Limnology.* New York: McGraw-Hill Book Company, 1963.

Westman, W. E., *Ecology, Impact Assessment, and Environmental Planning.* New York: John Wiley & Sons, Inc., 1985.

Wetzel, R. G., *Limnology.* New York: Harcourt Brace Jovanovich College Publishers, 1983.

What Water Shortage? New York: United Nations Environment Program (UNREP) (no date).

Where is Earth's Water Located? USGS water resources: http://wwwga.usgs.gov/edu/earthwherewater.html, pp. 1–2, February 2000.

Wiggins, G. B., "Trichoptera families." In *An Introduction to Aquatic Insects of North America,* 3rd. ed. Merritt, R. W. and Cummins, K. W., (eds). Dubuque, IA: Kendall/Hunt Publishing Company, 1996.

Wooton, A. *Insects of the World.* New York: Facts on File, Inc., 1984.

World Resouces 1986. New York: World Resources Institute (WRI) and International Institute of Environment and Development (IIED), 1986.

WRI and IIED. *World Resources 1988–89.* New York: World Resource Institute (WRI) and International Institute of Environment and Development (IIED), p. 133, 1988.

Index

abdomen, 3
abdominal, 129
abiotic factor, 3
acyclic cycle, 28
adaptations, 96–99
adaptive changes, 98
aeration, 3
aestivation, 3, 99, 101
alderflies, 123–124
allochthonous, 108
antennae, 3
anthropogenic pollution, 93
appendages, 3, 129
aquatic food chain, 38
aquatic organisms, 129
atmosphere, 27, 34
attachment to firm substrate, 97
aufwuchs, 78
autochthonous, 107
autotroph, 3, 11
autotrophic, 9

bacteria, 3
bare-rock succession, 58
bed load, 18
beetles, 118–119
benthic life, 99–100
benthic macroinvertebrates, 103–132
benthic macroinvertebrate biotic index, 153
benthic zone, 4, 76, 90
benthos, 77–78, 86, 90

biochemical cycles, 10–11, 25–35
  biological sampling, 189–256
  location, 192–195
  planning, 190–192
biochemical oxygen demand (BOD), 4, 163, 180
bioelements, 25
biogenic salts, 85
biomonitoring, 104, 149–154
biosphere, 8
biotic index, 4, 151–154
body adaptation, 97
bristles, 4
bulbous, 4
burrowers, 107

caddisfly, 95, 115–117
calcium carbonate, 4
carbonaceous BOD, 186
carnivorous, 4
carrying capacity, 52
channel alteration, 204
channel flow status, 206
chemical toxins, 141
chemically stratified lakes, 89
chlorophyll, 4
clarity, 4
Clean Water Act (CWA), 134
clean water zone, 164
climax community, 4
climbers, 107
clingers, 107

cobble, 192
community, 4
community ecology, 49
competition, 4
consumers, 11, 86
cultural eutrophication, 87

damselflies, 124–127
data, 4
decomposers, 11, 86
decomposition, 4
deep ancient lakes, 88
deposition, 19
desert salt lakes, 88
detritivore, 130
digestive tracks, 4
disease-causing agents, 137
dissolved load, 18
dissolved oxygen (DO), 5, 130, 142–144, 147, 163
  measurement of, 180–181
  monitor for, 214
dissolved solids, 144–145, 148
distribution, 51–52
divers, 107
dobsonflies, 96, 123–124
dragonflies, 124–127
drainage basin, 5
drift, 5, 109
drift net, 221

ecological equivalency, 48
ecological pyramids, 41–42, 45
ecology, 1–3, 11
ecosystem, 5, 8, 11
effluent stream, 17
elytron, 130
embeddedness, 204
emigration, 5, 51
energy, 45
energy flow, 37–45
entraining velocity, 18
entrainment, 18
environment, 11
environmental carrying capacity, 52

epilimnion, 81
establishment of a stream, 94
Euglena, 86
euphotic, 76, 90
eutrophic lakes, 88
eutrophication, 5, 87
evapotranspiration, 16, 23

fall turnover, 81
fecal coliform bacteria, 144, 148
filament, 5
First Law of Ecology, 63
First Law of Thermodynamics, 45
fishkill, 5
floodplain, 5, 23
food chain efficiency, 39–40
foraging, 5
freshwater ecology, 73–90
fry, 5
fungi, 5

gaining stream, 17, 23
gaseous cycle, 26
gill tufts, 5
gills, 5, 130
grid sampling, 193–195
gross primary productivity, 43
habitat, 5, 8, 95

heavy metals, 141–142
Henry's Law, 171
herbivore, 130
heterotroph, 5, 11
heterotrophic, 9
high turbidity, 91
homeostasis, 5, 9
hooks and suckers, 97
hydrologic cycle, 70
hydrosphere, 27, 34
hypolimnion, 81
hyporheal zone, 5

immigration, 5
impoundments, 89
infiltration capacity, 16, 23
influent stream, 17

inorganic plant nutrients, 137
intermittent stream, 16
invertebrates, 5

kick screen, 221

lake turnover, 81
lakes classification, 87–89
laminar flow, 18, 23
landscape ecology, 49
larva, 5
limiting factor, 5
leaf processing in streams, 108–109
leeches, 127–128
lentic, 74
lentic class, 90
lentic communities, 86
lentic habitat, 74–76
Lieberg Law of the Minimum, 79
limiting factors, 9, 79–86
limnetic zone, 76, 90
limnology, 74, 90
lithosphere, 27, 34
littoral zone, 75, 90
locomotion, 6
losing stream, 17, 23
lotic, 74
lotic habitat, 77–78

macroinvertebrates, 6
  sampling equipment for, 198–199
macrophytes, 6
marl lakes, 89
mature streams, 94
maximum sustainable yield value, 94
mayflies, 96, 111–113
mean, 195
meandering, 21, 24
median, 195
mesotrophic lakes, 87
metamorphic, 130
microhabitats, 6, 95
midges, 109, 117
mine drainage, 139
mode of existence, 106

modes, 195
mortality, 50
muddy-bottom sampling, 207–211

natality, 50
natural stream purification, 161
nektons, 78
net primary productivity, 43
neustons, 78, 91
niche, 6, 9
nitrate, 145–146
nitrification, 30
nitrogen, 35
nitrogen cycle, 29–31
nitrogen-fixation, 6
nitrogenous phase, 186
non-point pollution, 6
non-point source, 159–161
nutrient, 6, 91
nutrient cycles, 27

old field succession, 61
old streams, 94
oligotrophic lakes, 87
operculum, 130
organic, 6
organic chemicals, 137
oxygen demand, 147
oxygen-demanding wastes, 137
oxygen sag, 169–170
oxygen sag curve, 169–170

palp, 130
parasites, 7
pelagic, 91
pelagic organisms, 78
perennial streams, 16, 23
periphytons, 78, 86, 90
pH, 148
phosphorus cycle, 32–33, 35
photosynthesis, 34
photosynthetic rate, 91
physiology, 49
piercers, 106
pioneer community, 58
piscicide, 7

planktonic, 146
plankton sampler, 216–218
plankton tow, 231
planktons, 78, 86, 90
point source, 7, 159
polar lakes, 89
pollution, 7
pools, 23
population, 7
population controlling factors, 54–55
population density, 50, 63
population ecology, 47–63
population genetics, 49
population growth, 52–55
population system, 47–48
positive rheotaxis, 97, 100
positive thigmotaxis, 97, 100
post-sampling, 213
predators, 106, 130
primary succession, 58
producers, 86
productivity, 42–44
profundal zone, 76, 90
proleg, 130
pupa, 130

range, 195
recovery zone, 166–167
respirator, 7
riffle beetle, 119–120
riffle zone, 77
riffles, 7, 23–24
riparian corridor, 7
riparian vegetation zone, 206
riprap, 7
runoff, 7, 91

saltation, 19
sample collection, 198–213
sampling devices, 213–219
sampling nets, 214–215
sand, 193
saprobity of a stream, 161–167
scrapers, 106
Secchi disk, 218
Second Law of Thermodynamics, 45

secondary productivity, 43
secondary succession, 61
sediment deposition, 205
sediment or suspended matter, 137
sediment samplers, 215–216
sedimentary cycle, 26
self-purification, 157–186
septic zone, 166
Shelford's Law of Tolerance, 79
shredders, 106
sinuosity, 20–21, 24
  index of, 21
skaters, 106
snags, 192
snails, 127, 129
solid load, 18
species, 7
species diversity, 55–58
specific adaptation, 98–99
sprawlers, 107
spring overturn, 83
statistical analysis, 195
statistical concepts, 195–198
stenothermal, 81
stoneflies, 96, 114–115
stratification, 91
stratified lake, 91
stream competence, 19
stream pollution, 122–147
stream profiles, 20
stream purification, 181–186
stream water, 65–70
Streeter-Phelps equation, 184–185
submerged macrophytes, 193
substrate, 7
Surber sampler, 221
succession, 7
sulfur cycle, 33–35
suspended load, 18–19
suspended sediment, 147–147
swimmers, 107
symbiosis, 7
system ecology, 49

taxonomic hierarchy, 109
taxonomy, 109

temperature monitor, 214
terrestrial organisms, 129
thalweg, 21, 24
thermal pollution, 140–141, 148
thermocline, 81
Thienemann's Principles, 110
TMDL Rule, 135–136, 148
total phosphate, 146
traction load, 19
transect sampling, 193
transpiration, 68
trophic groups, 106
trophic level, 8
true flies, 96, 117–118
tubifex worms, 127
turbidity, 84
turbulent flow, 18, 24

ultimate carrying capacity, 52, 63
unidirectional flow of energy, 38

vegetated banks, 192–193
volcanic lakes, 88

wash load, 18–19
wastewater treatment, 171–172
water appearance, 203
water cycle, 8, 68
water movements, 91
water odor, 203
water penny, 120–121
water-soluble inorganic chemicals, 137
water strider, 122–123
whirligig beetle, 121–122
winter kill of fish, 83
winter stratification, 83
worms, 127

young streams, 94

zone of recent pollution, 165